GLOBAL ECOLOGY
Towards a Science of the Biosphere

D1455622

GLOBAL ECOLOGY
Towards a Science of the Biosphere

Edited by

Mitchell B. Rambler
ORI Inc.
Rockville, Maryland

Lynn Margulis
René Fester
Department of Botany
University of Massachusetts
Amherst, Massachusetts

ACADEMIC PRESS, INC.
Harcourt Brace Jovanovich, Publishers
Boston   San Diego   New York
Berkeley   London   Sydney
Tokyo   Toronto

ACADEMIC PRESS, INC.
1250 Sixth Avenue, San Diego, CA 92101

United Kingdom Edition published by
ACADEMIC PRESS INC. (LONDON) LTD.
24–28 Oval Road, London NW1 7DX

Library of Congress Cataloging-in-Publication Data
Global ecology : towards a science of the biosphere / edited by
  Mitchell B. Rambler, Lynn Margulis, René Fester.
      p. cm.
    Bibliography: p.
    Includes index.
    ISBN 0-12-576890-7
    1. Ecology. 2. Biosphere. I. Rambler, M. B. (Mitchell B.)
  II. Margulis, Lynn. Date- .   III. Fester, René. Date-
  QH541.G545 1988
  574.5—dc19   88-18087
                                                      CIP

Printed in the United States of America
89 90 91 92 9 8 7 6 5 4 3 2 1

CONTENTS

CONTRIBUTORS

Numbers in parentheses refer to the pages on which the authors' contributions begin.

Daniel B. Botkin (31), *Department of Environmental Studies, University of California, Santa Barbara, California 93106*

Michael J. Cosentino (75), *Sun Microsystems Inc., 2550 Garcia Avenue, Mountain View, California 94043*

Minoo N. Dastoor (31), *JPL-NASA, California Institute of Technology, Oak Grove, Pasadena, California 91125*

John E. Estes (75), *Department of Geography, University of California, Santa Barbara, California 93106*

M. Patricia Gildea (113), *Complex Systems Research Center, Institute for the Study of Earth, Oceans and Space, University of New Hampshire, Durham, New Hampshire 03824*

Joel S. Levine (51), *Atmospheric Sciences Division, NASA Langley Research Center, Hampton, Virginia 23665*

J. E. Lovelock (1), *Coombe Mill, St. Giles on the Heath, Launceston, Cornwall PL15 9RY, England, United Kingdom*

Lynn Margulis (1, 143), *Department of Botany, University of Massachusetts, Amherst, Massachusetts 01003*

Jerry M. Melillo (113), *NASA Headquarters, 600 Maryland Avenue SW, Washington, DC 20546*

Berrien Moore III (113), *Complex Systems Research Center, Institute for the Study of Earth, Oceans and Space, University of New Hampshire, Durham, New Hampshire 03824*

Bruce J. Peterson (113), *Ecosystems Center, Marine Biological Laboratory, Woods Hole, Massachusetts 02543*

Mitchell B. Rambler (143), *ORI, Inc., 1375 Piccard Drive, Rockville, Maryland 20850*

Edward Rastetter (113), *Ecosystems Center, Marine Biological Laboratory, Woods Hole, Massachusetts 02543*

David L. Skole (113), *Complex Systems Research Center, Institute for the Study of Earth, Oceans and Space, University of New Hampshire, Durham, New Hampshire 03824*

Paul A. Steudler (113), *Ecosystems Center, Marine Biological Laboratory, Woods Hole, Massachusetts 02543*

John F. Stolz (31), *Department of Biochemistry, University of Massachusetts, Amherst, Massachusetts 01003*

Charles J. Vorosmarty (113), *Complex Systems Research Center, Institute for the Study of Earth, Oceans and Space, University of New Hampshire, Durham, New Hampshire 03824*

INTRODUCTION

One of the greatest scientific challenges we face in the coming decades is to understand the fundamental nature of the system that supports life on our planet. This system, which encompasses all life and the space inhabited by it, is called the biosphere. The biosphere is a complex interplay of inextricably linked biological, chemical, and physical processes. These processes are integrated over space and time at the Earth's surface by interactions of the atmosphere, sediments, waters, lands, and biota. All life depends on the continuation of this interaction. Essential nutrients must be supplied to all organisms at adequate rates and concentrations, and all wastes must be removed. Supply and removal throughout the entire biosphere are largely the consequence of biological activity. Often the supplier and the remover are different types of organisms.

Although understanding the dynamics of the linkages that exist between the vast resources of the globe is extremely difficult, it is crucial to the formation of a science of the biosphere. We are beginning to see more and more scientific activity with this goal in mind (Fyfe, 1985; Botkin and Orio, 1986[1]). Human activities have had a great influence on the nonhuman pro-

---

[1] References, which are indexed, follow each of the seven sections in this book. Key general references to the emerging field of global ecology, even if not specifically referred to, are listed at the end of the introduction (p. xii).

cesses of the environment and will no doubt continue to do so. Human influence has already altered biospheric distribution in unprecedented ways. As the world population of *Homo sapiens* continues its exponential rise, our activities increasingly affect the resources upon which our lives depend. To recognize fully the present and future consequences of human activities, a science of the biosphere is urgently required.

Understanding the resilience and response of the biota to environmental perturbations requires knowledge of the interaction between the biotic and nonbiotic components of the Earth's surface. The science forged here, one that considers the Earth as a single system with global dimensions, faces enormous difficulties. Acquisition of new knowledge requires a highly integrated multidisciplinary approach.

Science of the late 1980s is no longer constrained by the lack of effective tools for biospheric study. Satellite remote sensing can now provide global coverage, a minimum requirement for biospheric introspection. Computer hardware and software advances made within the past decade can handle, store, and process prodigious quantities of data. Sophisticated analytical techniques and theoretical models provide a means of integrating the data, converting them to useful information. Used effectively, such tools may yield an understanding of the changing biospheric dynamic.

This book describes the incipient principles of global ecology. Traditional academic disciplinary boundaries will be trespassed. Aspects of global feedback mechanisms, atmospheric and oceanic circulation, productivity and biogeochemical cycling are discussed in the context of the interaction of the biota—the sum of all life—and its environment. We hope our biologically oriented readers will begin to appreciate the technological advancements in satellite and aircraft remote sensing that can help in understanding the extent, rate, and significance of environmental changes and perturbations. The influence of the Earth's biota on its global environment is a remarkable aspect of comparative planetology. The sum of the evolution of ecosystems becomes global evolution, and we now stand on the threshold of a new integration of knowledge about the surface of our planet. Imagery from space influences us to think of the fragments of details from different disciplines as a coherent whole, one that is becoming the science of the biosphere (Schneider and Londer, 1984) and is generating a great deal of public interest as well (Myers, 1984; Snyder, 1985; Southwick, 1985).

By way of introduction to our text, we show the structure of the Earth's atmosphere in relation to the chapters of this book (Figure i-1). Professor Levine's chapter reviews the structure and composition of the air mass

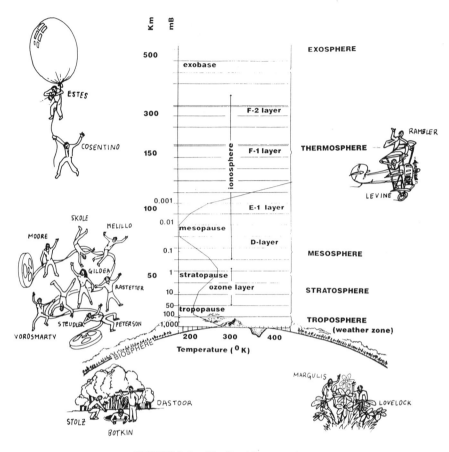

FIGURE i–1. The Earth's atmosphere.

blanketing the Earth (pp. 51–74). Professor Estes and coauthor Cosentino, from the vantage point of their stratospheric balloon, inform us about the radical shifts in perspective attributable to the new technologies of remote sensing and the space program (pp. 75–111). The potential unification of the sciences of the surface of the Earth (into the science of geognosy) is emphasized by Margulis and Lovelock, who encourage the use of the Gaia hypothesis as an organizing principle for the vast amounts of often disparate data that are rapidly accumulating (pp. 1–30). Our current knowledge of the biosphere, the place where life exists, is stressed by Stolz, Botkin, and Dastoor (pp. 31–49), whereas the ways in which the major biospheric geochemical (i.e., planetary metabolic) processes are assembled for study are discussed in Moore et al. (pp. 113–141). It is shown, in Rambler and Margulis (pp. 143–147), how the image of Earth from space has begun to inspire integrated study of the ecology of the entire planet by scientists from vastly different academic traditions. Together these chapters provide the prerequisites to the eventual standard college textbook that is yet to be written. Beyond the parochialism of current ecology, biogeochemistry, geology, meteorology, and other traditional codified academic disciplines, lies the increasingly real possibility of a new science of the Earth's surface.

REFERENCES

Dastoor, M. N., L. Margulis, and K. H. Nealson, eds. 1979. *Interaction of the Biota with the Atmosphere and Sediments.* NASA, Washington, DC.

Fyfe, W. S. 1985. The international geosphere-biosphere program: Global change. In: *Global Change*, T. F. Malone and J. G. Roederer, eds. The proceedings of a symposium sponsored by the International Council of Scientific Unions (ICSU) during its 20th General Assembly in Ottawa, Canada, on 25 September 1984. Pp. 500–508. Cambridge University Press, Cambridge.

Malone, T. F. and J. D. Roederer, eds. 1985. *Global Change.* Cambridge University Press, Cambridge.

Myers, N. 1984. *Gaia: An Atlas of Planet Management.* Anchor Press/Doubleday and Company, Garden City, New York.

Orio, A. A. and D. B. Botkin, eds. 1986. "Man's Role in Changing the Global Environment," Part A. The Science of the Total Environment, **55**:vii–399. Elsevier Science Publishers, Amsterdam.

Orio, A. A. and D. B. Botkin, eds. 1986. "Man's Role in Changing the Global Environment," Part B. The Science of the Total Environment, **56**:vii–415. Elsevier Science Publishers, Amsterdam.

Sagan, D., ed. 1985. *The Global Sulfur Cycle.* NASA Technical Memorandum 87570. NASA, Washington, DC.

Schneider, S. H. and R. Londer. 1984. *The Coevolution of Climate and Life.* Sierra Books, San Francisco.

Snyder, T. P., ed. 1985. *The Biosphere Catalogue.* Synergetic Press, London and Fort Worth.

Southwick, C. H., ed. 1985. *Global Ecology.* Sinauer Associates, Sunderland, Massachusetts.

Vernadsky, V. 1986. *The Biosphere.* Synergetic Press, London. (Translated from the French, which was originally translated from the Russian.)

Walker, J. C. G. 1986. *Earth History.* Jones and Bartlett Publishers, Inc., Boston and Portola Valley.

I GAIA AND GEOGNOSY

L. MARGULIS AND J.E. LOVELOCK

WHAT IS THE GAIA HYPOTHESIS?

THE HYPOTHESIS

The Gaia hypothesis is a theory of the atmosphere and surface sediments of the planet Earth taken as a whole. This hypothesis, in its most general form, states that the temperature and composition of the Earth's surface are actively regulated by the sum of life on the planet—the biota. Major aspects of the Earth's surface are dynamically maintained in frantic stability. That is, as changes in the gas composition, temperature, or oxidation state are induced by astronomical, biological, or other perturbation, the biota responds to these changes by growth and metabolism. The biological responses, taken together, serve to ameliorate the changes. This regulation of the Earth's surface activities by the biota and for the biota has been in continuous existence since the earliest appearance of widespread life—for at least 3 billion years. This Gaian view is a radical departure from the former widely accepted concept of modern evolutionists that life on Earth is surrounded by and adapts to an essentially static environment. A Gaian view not only has important implications for understanding life's past but is also relevant to the design and interpretation of observations and experiments on present life (Lovelock, 1988).

The Gaia hypothesis has come of age: we can now celebrate more than

1

twenty-one years of publication of Gaia papers in the scientific literature (Table I–1). Although only one official scientific meeting (see p.18) has been held on the Gaia hypothesis, the concept has come under increasing use as an organizing principle in the study of carbon and sulfur cycling (Sagan, 1985; Sagan, 1986a, b).

The Gaia hypothesis, until now of interest only to certain interdisciplinary scientists, may some day provide a basis for a new kind of ecological science (Lovelock, 1982). To emphasize the integration aspects of this planetary science we suggest the term *geognosy* (from Greek, meaning knowledge of the Earth). We are discouraged by the use of concatenated terms such as biogeochemistry, geomicrobiology, chemical oceanography, and microbial ecology, which underscore the fragmentation of the scientific approaches. We find that the term *environmental science* assumes a passive environment and the term *ecology*, originally designating a subfield of biology, is too diffuse. It has been used politically, socially, and scientifically to such an extent that the meaning of ecology is no longer clear.

HISTORY

Recognition of the Biosphere and the Impetus of the Space Program

Although the term *biosphere* first appeared in 1875 in E. Suess's work on the geological structure of the Alps, it was V.I. Vernadsky (1863–1945) who developed the concept extensively and who, since 1911, used it in its modern sense (V.I. Vernadsky, *Works*, v. 1, p. 178 (in Russian); see A.V. Lapo, *Traces of Bygone Biospheres*, Mir Publishers, Moscow, English edition 1982). Vernadsky said (*La biosphere*, 1929),

> The biosphere is the envelope of life, i.e., the area of existence of living matter . . . the biosphere can be regarded as the area of the Earth's crust occupied by transformers, which convert cosmic radiation into effective terrestrial energy: electric, chemical, mechanical, thermal, etc.

Since Vernadsky, there has been a continuous tradition (called *biogeochemistry* or *geomicrobiology*) in the USSR and to a lesser extent in Europe (Krumbein, 1978; Fersman, 1973) that has recognized strong interactions among the surface sediments, hydrosphere, atmosphere, and living organisms. (For example, in *Introduction to Physical Geography*, M.M. Yermolaev states, "the biosphere is understood as being that part of the geographical envelope of the Earth, within the boundaries of which the physico-geographical conditions ensure the normal work of enzymes" [Lapo, p. 20].) However, in the English-speaking world, the first expression of the notion in scientific terms that we could find was an editorial of the *Scientific Amer-*

TABLE I-1 GAIA HYPOTHESIS COMES OF AGE: TWENTY-THREE YEARS OF SCIENTIFIC PUBLICATIONS ON THE GAIA HYPOTHESIS (1965–1988)

Date	Authors	Title	Journal & Reference
1965	Lovelock, J.E.	A physical basis for life detection experiments	*Nature* **207**:568–569
1967	Hitchcock, D.R. and Lovelock, J.E.	Life detection by atmospheric analysis	*Icarus* **7**:149–159
1969	Lovelock, J.E. and Griffen, C.E.	Planetary atmospheres: Compositional and other changes associated with the presence of life	In: *Advanced Space Experiments,* Vol. 25, D.L. Tiffany and E. Galtzeff, eds. American Astronomical Society, Washington, DC
1972	Lovelock, J.E. and Lodge, J.P.	Oxygen—the contemporary atmosphere	*Atmospheric Environment* **6**:575–578
1972	Lovelock, J.E.	Gaia as seen through the atmosphere	*Atmospheric Environment* **6**:579–580
1974	Lovelock, J.E. and Margulis, L.	Atmospheric homeostasis, by and for the biosphere: The Gaia hypothesis	*Tellus* **26**:1–10
1974	Margulis, L. and Lovelock, J.E.	Biological modulation of the Earth's atmosphere	*Icarus* **21**:471–489
1975	Margulis, L. and Lovelock, J.E.	The atmosphere as circulatory system of the biosphere—The Gaia Hypothesis	*CoEvolution Quarterly* **6**:30–41
1975	Lovelock, J.E.	Thermodynamics and the recognition of alien biospheres	*Proc. R. Soc. Lond.* B **189**:167–181
1977	Margulis, L. and Lovelock, J.E.	The view from Mars and Venus	*The Sciences* Mar./Apr. pp. 10–13
1978	Watson, A., Lovelock, J.E., and Margulis, L.	Methanogenesis, fires, and the regulation of atmospheric oxygen	*BioSystems* **10**:293–298
1980	Margulis, L.	After Viking: Life on Earth	*The Sciences* Nov., pp. 24–26
1980	Margulis, L. and Lovelock, J.E.	L'atmosphère est-elle le système circulatoire de la biosphère? L'hypothèse Gaia	*CoEvolution* **1**:20–31

TABLE I-1 GAIA HYPOTHESIS COMES OF AGE (*continued*)

Date	Authors	Title	Journal & Reference
1981	Dastoor, M., Nealson, K.H., and Margulis, L., eds.	*Interaction of the biota with the atmosphere and sediments*	NASA Workshop Report, meeting Oct. 18–19, 1979, Washington, DC
1981	Doolittle, W.F.	Is nature really motherly?	*CoEvolution Quarterly* **29**:58–63
1981	Margulis, L., Nealson, K.H., and Taylor, I., eds.	*Planetary Biology and Microbial Ecology: Biochemistry of carbon and early life*	NASA Technical Memorandum 86043 (Summer program research report, 1982)
1982	Lovelock, J.E.	*Gaia: A New Look at Life on Earth*	Oxford University Press, Oxford and New York
1982	Lovelock, J.E. and Watson, A.J.	The regulation of carbon dioxide and climate: Gaia and geochemistry	*Journal of Planetary Science* **30**:795–802
1982	Margulis, L.	The biological point of view: The effect of life on the planet	In: *Formation of Planetary Systems*, A. Brahic, ed. Centre d'Etudes Spatiales, Capaudes-Editions, Toulouse, pp. 891–893
1983	Lovelock, J.E.	Gaia as seen through the atmosphere	In: *Biomineralization and Biological Metal Accumulation*, P. Westbroek and E. de Jong, eds., Reidel Publishing Co., Dordrecht, Holland, pp. 15–25
1983	Margulis, L. and Lovelock, J.E.	Le petit monde des pâquerettes: Un modèle quantitatif de Gaia	*CoEvolution* **11**:48–52
1983	Margulis, L. and Stolz, J.	Microbial systematics and a Gaian view of the sediments	In: *Biomineralization and Biological Metal Accumulation*, P. Westbroek and E. de Jong, eds., Reidel Publishing Co., Dordrecht, Holland, pp. 27–53
1983	Sagan, D. and Margulis, L.	The Gaian perspective of ecology	*The Ecologist* **13**:160–167
1983	Watson, A. and Lovelock, J.E.	Biological homeostasis of the global environment: The parable of Daisyworld	*Tellus* **35B**:284–289

TABLE I-1   GAIA HYPOTHESIS COMES OF AGE (*continued*)

Date	Authors	Title	Journal & Reference
1984	Sagan, D. and Margulis, L.	Gaia and philosophy	In: *On Nature,* Leroy Rouner, ed., **6**:100–125. Boston University Studies in Philosophy and Religion, University of Notre Dame Press, Notre Dame, Indiana
1985	Sagan, D., ed.	*Planetary biology and microbial ecology: The global sulfur cycle*	NASA Technical Memorandum (Summer program research report, NASA Ames, Jun.–Aug. 1984)
1986	Margulis, L., Lopez Baluja, L., Awramik, S.M., and Sagan, D.	Community living long before man	In: *Man's Effect on the Global Environment,* D. Botkin and A. A. Orio, eds. Vol. 2, Elsevier Science Publishers, Amsterdam; *The Science of the Total Environment* **56**:379–397
1987	Lovelock, J.E.	*Gaia: A New Look at Life on Earth,* 2nd ed. Oxford	University Press, Oxford and New York
1988	Lovelock, J.E.	*The Ages of Gaia*	W. W. Norton Co., New York

ican of June 1875. Here it was stated that the facts of nature were more consistent with life having made the Earth what it is than with the Earth having been created as a fit place for life.

Several scientists (including ecologist G.E. Hutchinson (1954), geologist H.A. Lowenstam (1974), the physical chemists G.N. Lewis and L.G. Sillen, oceanographer A. Redfield, and zoologist J.Z. Young), questioning the conventional wisdom about the geochemical and geophysical evolution of the Earth, have recognized some participation by life in the evolution of its environment. However, the concept of biologically driven environmental evolution has been generally ignored by formally trained geologists and geochemists.

Most geologists who model the evolution of the Earth ignore the intensely interactive influence of the biota. The counterpart of this scientific "apartheid" has been the failure of biologists to recognize that the evolution of the species is strongly coupled with the evolution of their environment. For example, in a collection of essays by distinguished biologists in which the editor J. Maynard-Smith (1982) brings together the most controversial issues of evolutionary biology, the sole mention of the environment comes in an essay by S.J. Gould, who concludes,

> Organisms are not billiard balls, struck in a deterministic fashion by the cue of natural selection and rolling to optimal positions on life's table. They influence their own destiny in interesting and complex, comprehensible ways. We must put this concept of organism back into evolutionary biology.

The Gaian view is that the biota and its environment constitute a single homeostatic system that opposes changes unfavorable for life. This view would probably have been kept like Cinderella in the scullery had not the National Aeronautics and Space Administration (NASA), in the role of the prince, offered a rescue by way of the planetary exploration program. When the Earth was first seen from outside and compared as a whole planet with its lifeless neighbors, Mars and Venus, it was impossible to ignore the overwhelming sense that the Earth was a strange and beautiful anomaly. Its evolution could not be explained solely in terms of conventional biology or geology.

The new questions raised by space science were largely technical, such as: How is life on another planet to be recognized? Because this question could not be answered by conventional science, it forced a new look at life on Earth and raised a second question: What are the differences between a planet, like the Earth, that bears life, and its neighbors Mars and Venus, which do not?

The most immediate and obvious difference among the Earth, Mars, and Venus is the composition of their atmospheres. The atmospheres of Mars and Venus are close to the chemical equilibrium state. The disequilibria that are present can be readily accounted for in terms of physical chemistry: interactions of solar radiation with the gas mixture. The Earth by contrast is anomalous in almost every aspect. Oxidizing and reducing gases coexist; carbon dioxide is present but at a concentration far below the expectations of chemistry. Unstable gases such as nitrous oxide and methane enhance the air with their unexpected presence.

The intense disequilibrium of the Earth's atmosphere reveals the presence of life. The persistence of this unstable atmosphere at a constant composition over periods vastly longer than the residence times of its constituent gases reveals the presence of a control system.

It is often difficult to recognize the larger entity of which we are a part; as the saying goes, "You can't see the forest for the trees." So it was with the Earth before we shared vicariously with astronauts that stunning and awesome vision—that perfect sphere that punctuates the division of the past from the present.

The new view of the Earth provided by NASA was not just aesthetic. Our planet and its near neighbors, seen with the more discerning instruments aboard spacecraft, generated curiosity. Prominent among the questions asked about the planets was: Do they bear life? Few doubted the importance of this question. In the words of Norman Horowitz (1974), "The discovery of life elsewhere would be a momentous event and enlarge our view of the universe and of ourselves." (See Horowitz, 1986, for the story.) Nearly all space scientists believed the planet most likely to bear life was Mars, and that the best way to find it was to send a spacecraft there equipped with an automatic biological and chemical laboratory.

But, according to the Gaia hypothesis, life detection did not require a planetary visit. Information about the composition of the planet's atmosphere was sufficient. It was argued that life on the planet would profoundly affect the atmosphere. On a planet such as Mars that lacks other fluid media—for example, open bodies of water—the atmosphere is a necessary mobile medium for the circulation of the elements required for life. The use of the atmosphere by life would change its composition to one that was recognizably different from that of the near-equilibrium state of a lifeless planet. On the basis of this argument and using the information then available from Earth-based infrared telescopes, Hitchcock and Lovelock (1967) concluded that Mars was probably lifeless.

To back up this rather unpopular conclusion, they used as evidence a gedanken experiment in which the Earth was viewed from Mars with an infrared telescope. Such a view of the Earth would easily have revealed the presence and abundance of the common gases O_2, CO_2, H_2O, CH_4, and N_2O. From this evidence and from a knowledge of the intensity of sunlight incident upon the Earth, it can be argued with near certainty that there is life or at least there are processes inexplicable by chemistry alone. Take, for example, two of the gases listed: oxygen at 21 percent and methane at 1.5 parts per million by volume (ppmv). It is not difficult to calculate from their probable rates of reaction that the observed steady state concentration requires an annual production of approximately 10^9 tons of methane and 4×10^9 tons of oxygen. No abiological processes make and sustain these two reactive gases in such quantities. It was reasoned, therefore, that they are probably the products of life.

In the late 1960s space scientists understandably rejected this approach. Not only did it suggest that the life detectors of the Viking Mission to Mars were otiose, but it also made public the fact that NASA was supplying funds for research to prove that there was life on Earth. But space science also generated this new look at the Earth, returning to questions posed earlier by the work of Hutchinson, Redfield, and Sillen. How is it that the Earth keeps so stable an atmospheric composition when it is made up of reactive gases? Still more puzzling: Why should so unstable an atmosphere be just right for life?

Thinking along these lines led to the idea that the atmosphere was more than just part of the environment—the atmosphere is regulated by life. Like the construction of a hornet's nest or a termite mound, the atmosphere is not living but rather an elaborately worked artifact of life that sustains a favorable environment. The novelist William Golding, to whom we are indebted, suggested the name Gaia, which the Greeks used for the goddess Earth, to label our hypothesis (Hughes, 1983).

THE ROLE OF MICROBES

The contribution of life in the active formation of its own environment has been suggested by isolated pieces of science. The Gaia hypothesis has tended to unite these, especially emphasizing the role of microbes.

The importance of microbes in weathering and the production of soil has generally been admitted (Alexander, 1977). The role of coccolithophorids, foraminiferans, and hydroids in the formation of calcium carbonate minerals has also long been recognized (Fersman, 1973; Lapo, 1982; Lowenstam, 1974). However, the concept that such activities form part of a planetwide regulatory system is still not accepted by mainstream science.

Whereas most microbiology in the Western world has been concerned with medicine, food, and agriculture, the tradition of environmental geomicrobiology in the Soviet Union has led to the formation of a body of literature and several research institutes that are not familiar to English-speaking ecologists, geologists, atmospheric chemists, and other environmental scientists. In a review of the Russian literature, A.V. Lapo (1982) claims,

> Among contemporary scientists writing in English there are two men—G.E. Hutchinson and H.A. Lowenstam—who can be said to have contributed most to the development of the ideas considered in this book (*Traces of Bygone Biospheres*). . . . G.E. Hutchinson began to be interested in problems of biogeochemistry by the early forties, under the influence of V.I. Vernadsky's son Georgi Vladmirovich.

Hutchinson's chapter in *The Earth as a Planet* (Kuiper, ed., 1954) provided an original and clear view of the influences of life on a planetary scale. His work on the *Biogeochemistry of Vertebrate Excretion* (1954) showed the importance of vertebrates in the accumulation of mineral deposits, such as guano. Hutchinson's many reviews and articles recognized the contributions of life to atmospheric gases (such as methane) and lithification (such as phosphorus-bearing rocks). H.A. Lowenstam has pioneered studies on biomineralization, the production of mineral substances inside living cells. In so doing, he has repeatedly emphasized the role of life in molding the environment (Lowenstam, 1974; 1989). For the most part, until recently (Westbroek and de Jong, 1983), scientists have worked independently, largely without institutional support for this sort of endeavor.

With the rise of molecular biology in western Europe and the United States (Judson, 1979) and comparative microbial physiology, the tools for the study of microbial evolution on the basis of comparisons of their macromolecules became available (molecular evolution). Phylogenies relating bacteria began to be devised (Margulis, 1969; Woese and Fox, 1977). Furthermore, direct studies of fossil morphological and chemical remains of microbes in well-dated pre-Phanerozoic sediments provided the basis for the realization that, prior to the evolution of animals and plants, over 80 percent of the history of the Earth was dominated by bacteria (Schopf et al., 1984). Widely divergent in their chemical and metabolic properties, all these microbes interact directly with the gases of the atmosphere. Indeed, most of the trace gases (methane, hydrocarbons, ammonia, carbon monoxide, and so forth) are released as direct products of microbial metabolism. These scientific developments conspired to produce the recognition of the importance of the microbial world in Gaian atmospheric regulation (Lovelock and Margulis, 1974; Margulis and Lovelock, 1974) and sedimentary processes (Margulis and Stolz, 1983).

In collaboration with our colleague A.J. Watson, we have tried to unify these disparate concepts. We now postulate the Gaian system as no more and no less than a colligative property of life. The phenomenon of Gaia as regulation arises automatically from the existence of physical and chemical constraints that limit the environment for life. Gaia emerges from the tendency for exponential growth when conditions are favorable and from natural selection. These notions were first expressed by Lovelock (1982) and have been developed subsequently by Watson and Lovelock (1983). Data, derived from recent observations of the diversity and abundance of microbial life in desiccated microbial mats, have been collected in support of this idea (Brown et al., 1985; Margulis et al., 1986).

GAIA, HOMEORRHESIS, AND CONTROL THEORY

Gaia as a Cybernetic System

Gaia, we maintain, is a cybernetic, or control, system. Cybernetic systems are "steered"; biological cybernetic systems are steered from the inside.

They actively maintain specified variables at relatively constant levels in spite of perturbing influences. Such systems are said to be homeostatic if their variables (such as temperature, direction traveled, pressure, light intensity, and so forth) are regulated around a fixed set point. Examples of such set points might be 22° C for a room thermostat or 40 percent relative humidity for a room humidifier. If the set point is not constant but changes with time, it is called an operating point. Systems that have operating points that change progressively instead of set points are said to be homeorrhetic rather than homeostatic.

Physiological regulatory systems usually exhibit homeostasis and homeorrhesis. This is well illustrated in the embryological examples described by C.H. Waddington (1976) where growth and morphogenesis are homeorrhetic, but the maintenance of steady states, such as those of pH and ionic strength, are homeostatic. Gaian regulatory systems, undoubtedly complex, multicomponented, and changing, must include homeorrhetic as well as homeostatic states.

Even minimal cybernetic systems have certain defining properties: a sensor, an input, a gain (the amount of amplification in the system), and an output. In order to achieve stability, the output is compared with the set or operating point such that errors are corrected. Error correction means that the output must in some way feed back to the sensor such that the new input can compensate for the change in output. Positive or negative feedback—usually both—are involved in error correction.

Cybernetic systems are elegant. They are not "mere mechanisms" with pieces that may be dissected by a reductionist. A wonderful aspect of cybernetic systems is the paradox of their imperfection. Such systems can only function because of errors. Their excellence is measured in terms of their ability to maintain, in the face of perturbations, a set goal such as temperature, blood pressure, or direction of travel. They can only approach, but never reach, perfection; their method is to accept a continuous state of error, and to sense it and oppose it as well as they can.

In a simple form of cybernetic system there is a goal to be achieved, a means of sensing any departure from error in the attainment of this goal (called error), a means of amplifying the error and of feeding it back as the means for its own opposition. This is the basis of most of the engineering control systems we encounter. Some biological systems also may work this way. However, more often they do not possess a single set goal, but rather a consensus among a number of associated systems that may cover a substantial range of possible states. There is no precise normal or fixed temperature in mammals, including people. Temperatures vary across our bodies according to our needs, and they vary through time as species evolve. When in a fever or when running, body temperature may be high; in starvation or at the extremities in cold weather, it may be lower.

Geologists accept that the Earth's surface features are a result of inter-action between the biota and the lithosphere, but they see this process as a passive one. Life adapts to the changes of its environment, and the com-position of the Earth's surface reflects the chemical transactions of the biota. Standard geochemical analyses require no active interaction between life and the environment. In sharp contrast, the Gaia hypothesis postulates a planet with the biota actively engaged in environmental reg-ulation and control on its own behalf.

Biologists have objected to this hypothesis on the grounds that it is Pan-glossian, nothing more than inflated optimism. Ford Doolittle (1981) claimed that the biota, in the normal pursuit of their local selfish interests, could not possibly have evolved an altruistic system global in scale. Dawk-ins rejects the Gaia hypothesis on the grounds that the emergence of Gaia requires "interplanetary selection." In order for Gaia to exist, Dawkins claims, "The universe would have to be full of dead planets whose homeo-static regulation systems had failed with, dotted around, a handful of suc-cessful, well-regulated planets of which Earth is one" (Dawkins, 1982).

These criticisms forced us to think again about the basis for Gaia. We knew that the evidence from the composition and the flux of gases through the atmosphere required the existence of a control system. Otherwise, in the Earth's unstable and reactive atmosphere it would not be possible to keep the observed steady state for periods long in comparison with the residence times of individual gases. How were we to reconcile the apparent fact of Gaia's existence with the sound theoretical argument, based on the properties of life, that it could not exist?

The Lesson of the World Without Daisies

These criticisms led to the formulation of a mathematical model to show how, in principle, the growth of organisms could affect the global temper-ature. In general, control systems that work well in nature are difficult to describe. For example, even a simple nonlinear control system, such as that of a laboratory oven controlled by a bimetallic strip thermostat, is difficult to analyze (Riggs, 1970).

The best compromise we have found so far for explaining the workings of a Gaian mechanism is a model, and shortly we will describe the regula-tion of the temperature of an imaginary planet over a wide range of solar radiation fluxes as a working example. But before we do, it is helpful to set the scene with an example of positive and negative feedback in a purely geophysical context so as to distinguish this limited form of regulation from that of the active regulation characteristic of Gaia.

Imagine a planet with a surface containing abundant water and consider how its average temperature will change as the flux of radiation from its star increases. Figure I–1 illustrates the course of this thermal evolution. For the sake of simplicity, the luminosity of the star is taken to increase linearly with time and the average surface temperature (T) of a planet with an albedo (A) would increase according to the relationship:

$$T = \left[\frac{F(1 - A)}{4s} \right]^{.25} \ldots \tag{1}$$

where (F) is the flux of radiation from the star and (s) is the Stephan Bolzman constant. This relationship is illustrated for a planet with a constant albedo, such as one with a dry, transparent atmosphere, by the dotted line. There is a monotonic increase of temperature with increasing stellar luminosity.

The temperature of a wet planet changes in a different way. At low luminosities the planet would be frozen with a high albedo. As long as it remained frozen, the average temperature would remain lower than that of the dry planet, although it would rise at the same rate as the radiation flux increased. This is illustrated on the figure by the solid line from A to B (Figure I–1).

Eventually the heat flux would be sufficient to melt enough ice to expose either bare rock or ocean. When this happens, the local albedo will fall and more heat will be absorbed, thereby increasing the area of the melted zone. This process, illustrated by the continuation of the solid line from B to C, will rapidly accelerate until a new thermal equilibrium is reached with most, if not all, of the planetary ice melted. From here onward, a further increase in heat flux increases the water vapor content of the planetary atmosphere and, through increasing cloudiness, could again increase the albedo. The temperature, as indicated by the solid line from C to D, would now rise less rapidly than on a dry planet. Eventually a constant cloudiness and albedo would become established, and the rise of temperature would resume the same rate as that for the dry planet, D to E.

This oversimplified example of the climatology of an imaginary planet admittedly ignores the greenhouse effect and many other processes affecting planetary surface temperature, yet it serves to illustrate passive positive and negative feedback. The positive feedback on temperature during ice melting is completely determined by the characteristics of water, as is the negative feedback of cloud cover. The melting of ice and the condensation

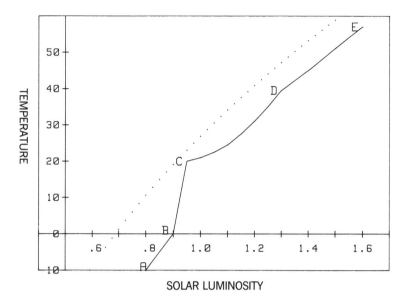

FIGURE I–1   Example of a daisy model calculation. Dotted line: increase in stellar (e.g., solar) luminosity as a function of age of star. The x-axis shows relative age of star (1.0 represents present, less than 1 represents past, and greater than 1 represents future). The y-axis is temperature in degrees Celsius. The temperature of a lifeless planet (e.g., a watery planet, with an unchanging albedo) would approximately reflect its sun's increase in temperature as seen by the solid line. (See text for further explanation.)

of water vapor to form clouds both take place at substantially constant and unchangeable temperatures. The passive regulation of planetary temperature by a purely geophysical process is limited to a small fixed range of temperatures. The regulation by the negative feedback is far from complete.

Other diagrams, e.g., Figure I–2, illustrate the active regulation of temperature on the imaginary planets by Gaian models. Regulation can be almost complete and cover a substantial range of radiation fluxes. Positive feedback is used to bring the planet swiftly to a temperature equable for the biota, and negative feedback is actively used to keep this goal accurate. The most important distinction between active and passive systems is their capacity for restoration after a perturbation. With an active Gaian system, this restoration tends to be rapid and complete over a wide range; with passive systems, it is only rapid or complete at some narrow and fixed range of values.

The Daisy World

Daisy world is an imaginary planet with a transparent atmosphere free from greenhouse gases and clouds; there is sufficient water present, however, to permit the growth of plants over nearly all the planetary surface. The dominant life forms are two species of daisy, one light in color and the other dark; there are also herbivores that feed on the two daisy species without reference and serve to recycle organic matter. Since they, like their background, are grey, the herbivores do not affect planetary albedo. Both light and dark daisies are identical in every respect other than color. Their growth rates vary with temperature in the same way. However, because they absorb more radiation, the local temperature of a stand of dark daisies will always be greater than that of a stand of light daisies. As a result, the rates of growth of the two species will be different at any given intensity of sunlight.

To model the evolution of this simple planetary ecosystem we shall further assume that the growth rate (β) of the daisies, like that of most plants, varies with temperature parabolically as follows:

$$\beta = A + BT - CT^2 \ldots \tag{2}$$

A, B, and C are constants chosen so that growth ceases below 5° C and above 40° C and is maximal at 22.5° C. Under cool conditions the dark daisies, which are locally warmer, will have the advantage of more rapid growth, whereas under hot conditions the cooler, light daisies will be favored. The results of competition for the planetary surface by the daisy species will be determined by the rate of spread of one into the zones of the other. We have used the relationship below, described and experimentally confirmed by Carter and Prince (1981), in our model:

$$\frac{dy}{dt} = \beta xy - \gamma y \ldots \tag{3}$$

where (x) is the number of sites susceptible to growth, (y) the number of infective sites, and (β and γ) the growth and death rates of the daisies, respectively.

Now let us examine the course of events as the luminosity of the sun that warms the daisy world increases in the manner of stars as they age. At first the planet warms to the same extent as would a lifeless planet, but when the surface temperature reaches 5° C, daisies will commence to grow. The growth of dark daisies will be favored and they will rapidly spread over the surface of the planet. As they spread, the planetary albedo will fall and what was at first a local warming of individual daisies becomes a global effect. This, plus the natural tendency for exponential growth until the supplies of energy and raw materials set a limit, exerts a powerful positive feedback on temperature. In the warmer climate, light-colored daisies now begin to grow and compete for space with the dark daisies. As a consequence, the temperature will not increase indefinitely since the spread of white daisies has a net cooling effect. As the stellar luminosity increases, so the area covered by light daisies enlarges until eventually the planet becomes too hot even for them. A catastrophic collapse of the eco-systems ensues as a result of rapid heating from the exposure of an ever-increasing area of bare and darker rock.

Figure I–2 illustrates the growth and decay of the populations of the two different colored daisies with increasing stellar luminosity. Also illustrated is the monotonic rise of temperature for a lifeless planet (dotted line) and the efficient thermostasis (solid line) attributable to the presence of the daisies.

Instead of two competing daisy types, we can model with one type, which has the capacity to adapt its color to that most comfortable for a given radiation intensity—rather as our skin colors vary with the intensity of the ultraviolet radiation to which they are exposed. Or, we can make color change dependent upon mutation followed by the natural selection of the type that grows best at the given heat flux. In both these examples, thermostasis is as strong and complete as in the example illustrated in Figure I–2.

The models illustrate how the powerful capacity of living systems to grow exponentially drives their environment to a state favorable for them. Living populations of organisms can act as an amplifier. Natural selection can serve as a sensor in a control system. In principle, such a system is able to maintain planetary surface temperature close to that favored by the plants—and it does so without any foresight or planning by the daisies. All that is required is their opportunistic local growth when conditions favor it.

This type of model is not intended to illustrate any known or imagined regulation processes at present in operation on the Earth. It was developed specifically to answer the objection by some biologists (Dawkins, 1982; Doolittle, 1981) that there was no possibility of natural selection leading to "altruism" on a global scale.

The daisy world model is not limited to the highly artificial conditions invented here. A more general version could take into account the possibility that once life appears on a planet it will be limited by the circumscribed set of physical and chemical constraints that characterize the biota.

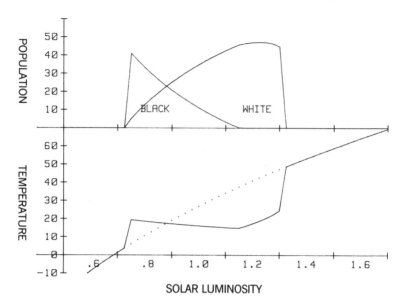

FIGURE I–2   Example of a daisy model calculation. The effect of black and white daisy growth, via albedo changes, on the modulation of the plant's surface temperature. Axes labeled as in Figure I–1. Black daisies, locally absorbing heat, grow to large population densities earlier in the planet's history when temperatures are colder (top graph). Dotted line indicates course of surface temperature expected on a lifeless planet, solid line indicates effect of daisy populations on surface temperature (See Watson and Lovelock, 1983, for details of these kinds of calculations).

Any internal or external change away from this set of conditions will lead not only to adaptation, which is the conventional wisdom, but also to the selection of those organisms whose growth alters the environment so as to oppose the unfavorable development. In addition, models of the regulation of atmospheric gases, such as oxygen or carbon dioxide, can be developed on a similar basis.

An elegant model of daisy world has been developed by C. Powys-Lybbe (1982, personal communication) in which the zonal variation of temperature and daisy growth are illustrated on an evolving series of world maps. The essential simplicity of the model can be displayed in animated form on a computer screen.

This form of dynamic model also provides insight on the way in which the evolution of the species is actively coupled with the evolution of the environment. Almost all previous discussion of evolution has taken place in the context of a fixed or, at most, an independently evolving environment (Maynard-Smith, 1982).

STATUS OF GAIA AND THE FUTURE OF OUR QUEST

ATMOSPHERIC GASES: EVIDENCE FOR GAIAN HOMEORRHESIS

Living organisms strongly interact with the oceans, the regolith (soil and other loose covering at the surface of the planet), the crustal rocks, and the atmosphere. Of these compartments, the least massive one, the atmosphere, is the most profoundly influenced by the biota. With the exception of phosphorus, all the elements required for the reproduction of living cells (C, H, N, O) flux through it. Not surprisingly, the strongest indications for the active regulation of chemical composition come from the atmosphere.

Figures I–3 and I–4 illustrate the abundances and fluxes of atmospheric gases for the present live Earth and compare them with those of a lifeless planet of the same chemical composition and distance from the sun.

The present atmosphere of Earth departs grossly from the expectations of the equilibrium of steady state, abiological chemistry. Indeed, it is like a dilute version of the fuel-air mixture drawn into an internal combustion engine: it is a reactive mixture. Hydrocarbons, methane, and other reduced gases react with oxygen. When illuminated by sunlight these gases react in a slow combination. Were they not continuously replaced, gases such as methane and hydrogen would vanish in a few decades. We know that the major sources of these gases are metabolic activities, primarily of bacteria (Margulis and Lovelock, 1974; Sagan and Margulis, 1983). Even the abundant nitrogen, as observed long ago by G.N. Lewis (Lewis and Randall, 1923), is not stable in the presence of oxygen or carbon dioxide on a watery planet. The stable compound of nitrogen on a lifeless Earth is the nitrate ion in solution in the oceans.

Carbon dioxide in the Earth's atmosphere is far less abundant than chemistry alone would permit. In the absence of life, the only significant sink for this gas is the reaction with calcium silicate as rocks are weathered; the rate of this process is limited by the slowness of the diffusion of CO_2 from the air to and within the soil. On the present Earth, atmospheric CO_2 is at least 30 times less abundant than would be expected of the abiological steady state (Lovelock and Watson, 1982). The vast perturbation of the Earth's atmosphere by life (i.e., decrease in carbon dioxide, increase in oxygen) is depicted in Figure I–5, which compares the present gaseous composition of these three inner planets.

The intense disequilibrium of our atmosphere advertises the presence of life. The maintenance of this reactive and unstable mixture at a constant composition over periods of time that are vast compared with the residential times of the individual gases is inexplicable by chemistry alone. It requires a regulated system that we call Gaia.

PARTIAL PRESSURES

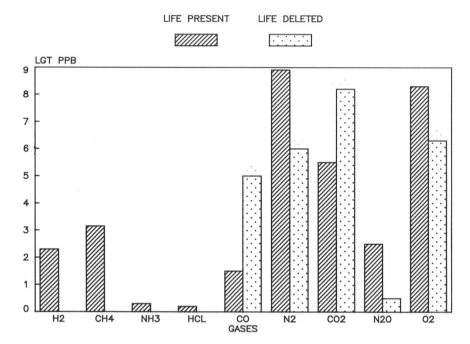

FIGURE I–3 Partial pressure of reactive gases of the Earth's atmosphere at present (with life) and those calculated to result from reactions were life to be extinguished. X-axis: H_2 hydrogen, CH_4 methane, NH_3 ammonia, HCl hydrogen chloride, CO carbon monoxide, N_2 nitrogen, CO_2 carbon dioxide, N_2O nitrous oxide, O_2 oxygen. Y-axis: gas concentration is log base 10 of quantities in parts per billion. Note how, after "life is deleted," all gases would reflect the tendencies of the atmospheres of the inner planets of the solar system to oxidize with time.

TESTS OF THE GAIA HYPOTHESIS

Chapman Conference

An entire Chapman Conference of the American Geophysical Union (AGU), held March 6 to 12, 1986, was devoted to formulating proofs of the Gaia hypothesis. See the proceedings[1] of this meeting for details and debate.

---

[1] Schneider, S., Boston, P. Chapman Conference on the Gaia Hypothesis, 1988. Manuscripts in preparation.

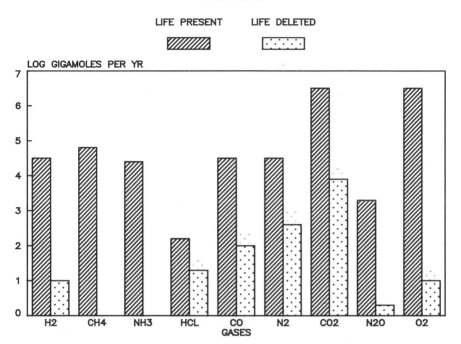

FIGURE I-4   Fluxes of the reactive gases of the Earth's atmosphere at present (with life) and those calculated to result if life were extinguished. X-axis: same as Figure I-3. Y-axis: Log base 10 of fluxes in 10^9 moles per year. Note the striking presence of immense methane and ammonia fluxes due to the metabolic processes of methanogenic and other organisms.

Global Perturbations

The Gaia hypothesis is testable and gives rise to predictions. We have postulated a control system, and there are several ways to test such a system. Among them are perturbation experiments; these are made by observing the controlled variable when some continuous or impulse changes are taking place. Liveliness and accuracy of regulation both suggest the presence of a control system and are a measure of its health. Reductionist experiments are usually inappropriate with control or living systems; little is learned of physiology from a dead or dissected animal.

The size of the system, the whole surface of the Earth, makes direct perturbation experiments expensive. Merely to label the whole of the least massive compartment, the atmosphere, with a tracer that can be detected

Planetary Atmospheres

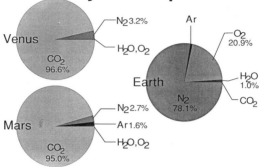

FIGURE I–5   Major gases of the atmosphere: Mars, Earth and Venus compared. The similarity of Mars to Venus relative to the Earth is apparent. Abbreviations as in Figure I–3. Ar (=argon) is somewhat less than 1% and CO_2 somewhat more than 0.03% on Earth. (Graphics courtesy of Jeremy Sagan and Business and Professional Systems, Cambridge, MA)

with exquisite sensitivity requires either a sizable nuclear explosion or the release of several tons of an expensive chemical, such as a heavy methane or a fluorinated carbon compound. Fortunately there is the alternative approach, that of using opportunistically accidental or inadvertent global perturbations. Among these are the consequences of the impact of an extraterrestrial body, such as a large meteorite or small planetesimal (see p. 22). Perhaps the present growth of the human population, with its increasing occupancy of the available habitats, may eventually constitute a perturbation great enough to measure. From the perspective of dramatic climatic changes discernible from study of the Earth's recent and fossil record so far, neither the accumulation of CO_2 nor the widespread changes in land use have discernibly altered the climate against the background of noise. Because of this noise level we cannot yet say whether the absence of detectable climate change is in itself a result of the presence of a control system.

Oxygen regulation

Oxygen is present in the atmosphere at 21 percent by volume. It has probably never risen above 25 percent nor fallen below 15 percent during the past 200 million years. The evidence for this limited range of oxygen concentration comes from the frequent presence of charcoal in the sedimentary column. This charcoal is a fossil record of ancient forest fires, and such fires can only occur if the oxygen is above 15 percent, since vegetation will not burn at less, and below 25 percent, since at greater levels fires are so fierce and all-consuming as to prevent the formation of forests. The atmospheric residence time of oxygen is about a thousand years. Therefore, the Earth has maintained an approximately constant concentration of oxygen for 200,000 times its residence time. Yet during pre-Phanerozoic times, probably in the late Archean Aeon, oxygen rose in concentration by approximately seven orders of magnitude (Cloud, 1974). These observations suggest the existence of a powerful system for oxygen regulation. Alternately, the upper limit of oxygen concentration may, by coincidence, be maintained at a favorable level by some geochemical limit.

The only net source of oxygen is the burial of carbonaceous material in the deep sediments, and the only sink is the reaction with reduced materials exposed during the weathering of rocks and by tectonic processes. Of the organic detritus deposited in the anaerobic sediments by the biosphere, only about 5 percent is in fact buried. The larger 95 percent is returned to the atmosphere as methane. All methanogenesis on the Earth today is a product of the metabolism of a peculiar group of anaerobic bacteria, the methanogens (Balch et al., 1979; Mah et al., 1977). Clearly a further rise in oxygen levels is not limited by a passive geochemical process but is determined by the active metabolism of methanogenic bacteria that directly influences the proportion of organic detritus that is buried.

The mechanisms of oxygen regulation are not fully known. Indeed, there are probably a number of interacting methods by which the O_2 is modulated (Garrels et al., 1976; Watson et al., 1978). R.L.M. Synge has speculated concerning a further possible mechanism. Among the products of the biota most resistant to microbial degradation are lignin, humic acids, and similar substances. These will be those most likely to be buried rather than returned to the atmosphere as methane. Synge proposed that the biogenesis of the precursors of lignins in plants might be oxygen sensitive. An increase in atmospheric oxygen would thus diminish the proportion of lignin made and consequently of carbon buried (Synge, 1983).

Carbon Dioxide Regulation

Carbon dioxide is the principal gaseous component of the atmospheres of Mars and Venus. It would also be a major component of the atmosphere of the Earth were it not for the presence of life. Carbon dioxide is present at about 340 ppmv. The geochemical considerations affecting the abundance of CO_2 may have been described by Garrels et al. (1976). The balance is determined by the fact that the only net source of CO_2 is tectonic. The only net sink is the reaction of CO_2 with calcium silicate during the

weathering of rocks (Garrels et al., 1976). In the absence of life the rate
of this reaction is slow. It would be even slower if the continents of a lifeless planet were dry, as the climate modeling of Shukla and Mintz (1982)
suggests. The present low levels of CO_2 are a consequence of biological
activities. The CO_2 content of the gases of the soil is between 10 and 40
times greater than the atmospheric abundance. By far the greatest quantity
of CO_2 is in the form of biogenic calcium carbonate rocks. A substantial
proportion of all of the carbon photosynthesized by the biota is actively
driven down into the sediments: growth of roots, descent of microbes with
draining water, and settling of dead plankton all provide examples. As the
ceaseless process of downward drift and burial proceeds, consumers at
every stage oxidize and convert organic compounds to CO_2. Through these
processes of settling and oxidation, and by direct biotic calcium carbonate
precipitation (Golubic, 1973), the biota draws CO_2 from the air into the
sediments and oceans.

Walker et al. (1981) proposed that a continuous decline of CO_2 abundance from an initial 30 percent by volume could explain the constancy of
the Earth's climate in spite of growth of luminosity of the sun as it aged.
This may be, but the controlling mechanism, the weathering of calcium
silicate by CO_2, is itself made possible by biological activity.

Meteoric Impacts: Perturbation and Recovery

The evidence from the atmosphere points to the existence of a control
system, Gaia, but the complexity of the couplings between the component
parts of the system, living and nonliving, seems to be too great for useful
modeling.

We are at a stage in planetary understanding similar to that of the early
physiologists in their attempts to comprehend animals. In such a state of
ignorance, the classic test of a postulated control system is to perturb it
and then observe the response.

It happens that approximately every seventy million years the Earth col-
lides with a planetesimal about ten kilometers in diameter; the occurrence
of these events is well reviewed by McCrae (1981) and Napier and Clube
(1979). The impulse energy of these impacts is in the range of 10^{20} joules.
This is an immense amount of energy: about 1,000 times greater than the
energy release would be of a total nuclear war in which the entire weapon
stock of the world was consumed. These perturbations are large enough
for their effects to be included in the geological record. Where the impacts
occurred on ancient continental rocks, visible craters of about 200 kilometers in diameter are produced.

Thin layers of rock rich in elements characteristic of extraterrestrial bodies have been discovered (Alvarez et al., 1980). One such layer is distributed worldwide at the boundary between the Cretaceous and the Tertiary periods. The boundary is visible and marks the extinction of some 60 percent of Cretaceous species of marine microorganisms.

We do not yet know the sequence of events at this time, but there seems little doubt that the interest stirred by the discoveries of Alvarez and his colleagues will encourage further geological exploration and new information. For a test of Gaia, it may be sufficient to know that 60 percent of marine organisms was destroyed about 65 million years ago, at the time marked by the boundary between the Cretaceous and the Tertiary. Whatever the cause, even assuming that the planetesimal played no part in it, such a deletion would by our hypothesis have large and long-lasting consequences. We have proposed that the atmospheric abundance of CO_2 is regulated and that part of the mechanism of regulation involves the removal of CO_2 by photosynthesis. The deletion of a sizable proportion of photosynthesizers would therefore be followed by a rise in CO_2 to a new steady level. This in turn would have led to a rise in global mean temperature to a level that would be less favorable than the preextinction regime. The new, more rigorous environment, forcing further extinctions, would have led to the evolution of other species. Eventually, the altered biota would reach its former vigor and restore climate and environment to their previous favorable state.

This scenario is one of several that could be imagined, and we await with interest the discoveries in the field that will confirm or deny it. If we look at the course of extinctions in this way, then we see that the time required to return to altered optima of temperature and atmospheric pressure might be very long indeed. However rapid the initial impact events, the consequences could reverberate for thousands or even millions of years. The time constraints of some geological processes are that long.

If we accept a geological perspective, assuming that the extinctions associated with the Cretaceous/Tertiary boundary are consequences of slow environmental changes as well as direct impact damage, then the distinction between "punctuated" (rapid) and "gradual" (slow) evolution seems to fade. If the environment and the species evolve in a coupled manner, evolution is constrained by the pace of environmental change.

Perturbation by planetesimal impact has been a regular feature of the Earth's history. As many as 30 major impacts have so far occurred. The major impact craters discovered on the ancient continental shield of Canada, where the record is better than elsewhere, are diagrammed in Figure I–6. The persistence of life and environment in spite of these frequent violent insults suggests the existence of Gaia as a vigorous, evolving cybernetic system.

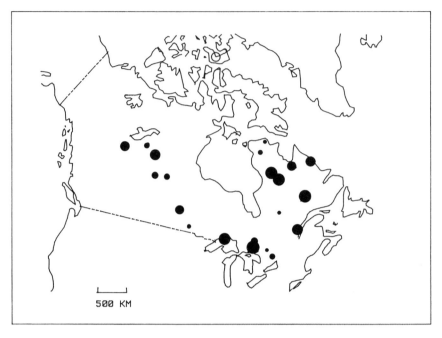

FIGURE I-6   Major meteoritic impact craters of Canada (solid circles).

EXPANSION OF GAIA TO MARS AND BEYOND

The notion of terraforming Mars, transforming it into a planet habitable for life, has been discussed often in science fiction. More recently it has been considered a possible future space project. Leaving aside any discussion of the moral questions raised by such planetary manipulation (costliness and potential to drain the Earth's resources, the alteration of the pristine, etc.) we might view the habitation of Mars as a splendid Gaian experiment. What might we do to Mars to enable it to become the second homeorrhetic biosphere orbiting the sun?

The present climate and surface composition of Mars is incapable of supporting our form of life. Brief growth or the mere survival of seeds or spores might be possible at the Martian equivalent of oases. It is important, though, to recognize that unless growth is sustained and expands to cover a substantial area of the planet, there is no way that the presence of life can maintain its environment. It is only when growth alters the local environment in its favor that expansion becomes possible. Then, through positive feedback, colonization proceeds until a substantial part of the planet is made habitable.

We have long believed, because of the absence of Gaian atmospheric and sedimentary inhomogeneities, that the search for sparse life at rare sites on Mars was pointless. Life must either be provided with or regulate its environment. Appropriate fluids, especially water, and gases or particles that cycle the elements required by living systems are mandatory. Required acidities, salinities, temperatures, and pressures must be maintained or, of course, life will perish. The timing of these biogeochemical cycles must be rapid relative to most geological processes in order to sustain the metabolism, growth, and reproduction of cells. Life is unlikely to take hold on a planet unless it has the capacity to cycle the appropriate elements and thereby modulate the climate and the atmospheric composition.

How could Mars be changed sufficiently to give propagules of life the chance to grow and flourish? The present Martian surface has the sunlight and chemical elements needed for the growth of microorganisms, especially CO_2 and nitrogen. There may be too little water and such as there is may be unavailable because of salinity and low temperatures. The major problems for life on Mars are lack of water and extreme variations of temperature around an unacceptably low average. To terraform Mars, therefore, the mean temperature must be raised and the size of the temperature fluctuations dampered. Sufficient water must be unfrozen from beneath the surface to provide for sustained growth once the planetary temperature is raised.

The terraforming of Mars could begin if a sufficient quantity of stable but strongly infrared-absorbing gas were provided. Such a gas would be necessary to provide a preliminary gaseous greenhouse. On Earth a small change in CO_2 can affect climate, but this is only very slightly a consequence of infrared absorption by CO_2 itself. The larger effect is due to the absorption of infrared radiation by water vapor. A small rise in temperature due to CO_2 is in effect amplified by the concomitant rise in water vapor content, assuming a constant relative humidity. In a similar way, if a greenhouse agent were sent to Mars it would have its effect multiplied by the accompanying rise of both CO_2 and water vapor in the Martian atmosphere. As the surface warmed, more of these gases would be released.

Promising chemicals for terraforming Mars might be the fluorocarbons. They intensely absorb infrared radiation in the 8- to 14-micrometer atmospheric windows. At the same time, they are photochemically inert and nontoxic. Their use for climate modification was proposed in a patent application by Lovelock in 1960. It has been calculated that a few thousand or tens of thousands of tons of CF_4 or a similar fluorocarbon would cause a sensible increase of the Martian surface temperature. In the thin atmosphere of Mars, the concentration of fluorocarbon would be about

10^{-8} by volume, sufficient by itself to provide some increase in surface temperature. This slight climatic amelioration would in turn be substantially amplified by the rise in CO_2 and water vapor that followed.

The effect of these greenhouse agents would persist only so long as they remained in the atmosphere. CF_4 is likely to have a residence time of tens of thousands of years, long enough to give a fair trial of the method.

The introduction of fluorocarbons to the Martian atmosphere of course is no more than the setting for a production. We are not equipped to model the changes of the Martian atmosphere and surface following the deliberate injection of fluorocarbons. Even if the greenhouse effect were achieved, the rate of transfer of heat to the ice locked in the permafrost might be unacceptably slow. But, if by any means the surface of Mars can be transformed into a seed bed, what shall we plant there to continue the process of planetary modification? Organisms associated with microbial mats from cold, dry regions of the Earth might be introduced to Martian habitats. Growing as dark-colored mats, the process of planetary warming might be further enhanced by the lowering of the planetary albedo. The microbial mats would act like the dark daisies of daisy world. The absence of oceans on Mars might actually favor biological climate control. The difference between the albedo of dark microbial mats and Martian soil is not great, but much of the planet might be opened for cover by microbial growth. By contrast, on Earth the greater part of the planetary surface is dark, unchangeable ocean. If growth covered a subtantial area of the planetary surface, microbial communities alone might effect albedo changes sufficient for climate regulation.

The crucial problem of Martian habitability is that of maintaining a sufficient quantity of water for life. Another major problem would be the replacement of CO_2 removed from the atmosphere for growth. The removal of CO_2 from the atmosphere would weaken the gaseous greenhouse effect. We proposed earlier here that on Earth this problem was resolved during the Archean Aeon by the production of methane by methanogenic bacteria. Martian planetary evolution might follow a similar course: a photochemical production of methane by methanogenic bacteria. The planetary atmosphere might even be anaerobic for a while. A photochemical production of polymeric organic substance might even provide a surrogate for the ozone layer. Such substances would reduce or remove the UV stress that the microbes would otherwise experience.

Even if this largest of all "big science" experiments—an experiment where biologists challenge physicists at their own game—is never attempted, it still is a fruitful source of ideas. If solar system habitability beyond the Earth's atmosphere is to be achieved, modeling is an essential first step as a part of the search for a potential new habitat for people. Here are a few of the questions generated by the concept of terraforming, a subject in great need of more detailed analysis:

1. How much fluorocarbon is needed and in what sort of timed introduction to effect enough rise in temperature sufficient to melt ice?
2. After the initial seeding of the Martian atmosphere with fluorocarbon, how will the atmosphere evolve? When? What are the optimal mean temperatures, pressures, and chemical composition that are achievable?
3. How long would it take to establish suitable microbial habitats and fill them with a relatively stable planetary biota? That is, can Gaia be inoculated onto Mars?

The transfer of Gaia, that is, the terraformation of any planet beyond the Earth, may be an impossible dream, unachievable even in principle. Whether or not Gaia can expand off the planet is an unsolved scientific problem. On the other hand, the perpetration of humans and human civilization on any planet in the absence of the complex Gaian support system is a delusion.

SUMMARY

This chapter provides a statement of the Gaia hypothesis which asserts that the biota regulates the major features of the planet (e.g., chemical composition of the atmosphere and surface temperature). Since work on the hypothesis has been scattered in the scientific literature since 1965, we tabulate here a comprehensive list of publications explicitly dealing with the Gaia concept. We review precedents and history of the Gaia concept as well as the importance of the space exploration program in its genesis.

We explain how the Gaian system, unlike any engineered one, is not controlled by a "steersman" or governor from the outside. Rather, like any living system, Gaian control is homeorrhetic. Set points change through time: the apparent stability is dynamically maintained by organization inside the system itself. A simple class of models (Daisy models) shows how, using ordinary concepts of population growth and physiology, dynamic environmental stability emerges as a consequence of the effect of growing and metabolizing organisms on their immediate environment. No special mechanisms of environmental control are invoked. We close by providing some testable predictions and indicating their impact on experimental and observational science in the acceleration of attempts to establish or reject the Gaia hypothesis.

ACKNOWLEDGMENTS

J.E. Lovelock acknowledges the Leverhulme Foundation, and L. Margulis acknowledges NASA Life Sciences (NGR004025), the Boston University Graduate School, and the Lounsbery Foundation in support of this work. Michael Dolan aided in the preparation of the manuscript.

REFERENCES

Alexander, M. 1977. *Introduction to Soil Microbiology*. John Wiley and Sons, New York.

Alvarez, W., L.W. Alvarez, F. Asaro, and H.V. Michel. 1980. Extraterrestrial cause for the Cretaceous-Tertiary extinction. *Science* **208**:1095–1108.

Balch, W.E., G.E. Fox, L.J. Magrum, C.R. Woese, and R.S. Wolfe. 1979. Methanogens: A reevaluation of a unique biological group. *Microbiological Reviews* **43**:260–296.

Brown, S., L. Margulis, S. Ibarra and D. Siqueiros. 1985. Desiccation resistance and contamination as mechanisms of Gaia. *BioSystems* **17**:337–360.

Carter, R.N. and S.D. Prince. 1981. Epidemic models used to explain biogeographical distribution limits. *Nature* **293**:644–645.

Cloud, P.E., Jr. 1974. Evolution of ecosystems. *American Scientist* **62**:54–66.

Cloud, P.E. Jr. 1983. The biosphere. *Scientific American* **249**: 176–189.

Dawkins, R. 1982. *The Extended Phenotype*. W.H. Freeman Co., San Francisco.

Doolittle, W.F. 1981. Is nature really motherly? *CoEvolution Quarterly* **29**:58–63.

Fersman, A.E. 1973. *Geoquimica Recreativa*. Editorial Mir, Moscow.

Garrels, R.M., A. Lerman, and F.T. MacKenzie. 1976. Controls of atmospheric oxygen: Past, present and future. *American Scientist* **61**:306–315.

Golubic, S. 1973. The relationship between blue-green algae and carbonate deposits. In: *Biology of the Blue-Green Algae*, N.G. Carr and B.A. Whitton, eds. Blackwell Scientific Publications, Oxford. Pp. 434–472.

Gould, S.J. 1983. In: *Evolution Now a Century after Darwin*. J. Maynard-Smith, ed. The Macmillan Press, London. Pp. 129–145.

Hitchcock, D.R. and J.E. Lovelock. 1967. Life detection by atmospheric analysis. *Icarus* **7**:149–159.

Horowitz, N.J. 1974. Oral communication.

Horowitz, N.J. 1986. *To Utopia and Back: The Search for Life in the Solar System*. W.H. Freeman, New York.

Hughes, J.D. 1983. Gaia: An ancient view of our planet. *The Ecologist: Journal of the Post Industrial Age* **13**:54–60.

Hutchinson, G.E. 1948. On living in the biosphere. *Scientific Monthly* **67**:393–397.

Hutchinson, G.E. 1954a. *Biogeochemistry of Vertebrate Excretion*, American Museum of Natural History, New York.

Hutchinson, G.E. 1954b. Biochemistry of the terrestrial atmosphere. In: *The Earth as a Planet*, G.E. Kuiper, ed. University of Chicago Press, Chicago. Pp. 371–433.

Judson, H.F. 1979. *The 8th Day of Creation: Makers of the Revolution in Biology*. Simon and Schuster, New York.

Krumbein, W.E. 1978. *Environmental Biogeochemistry and Geomicrobiology*. Vols. 1, 2, and 3. Ann Arbor Science Publishers, Ann Arbor, Michigan.

Kuiper, G.P., ed. 1954. *The Earth as a Planet.* University of Chicago Press, Chicago.

Lapo, V. 1982. *Traces of Bygone Biospheres* (published in Russian in 1979; English edition translated by V. Purto), Mir Publishers, Moscow.

Lewis, G.N. and M. Randall. 1923. *Thermodynamics.* McGraw-Hill Book Co., New York.

Lovelock, J.E. 1982. *Gaia: A New Look at Life on Earth.* Oxford University Press, Oxford and New York.

Lovelock, J.E. 1983. Gaia as seen through the atmosphere. In: *The Fourth International Symposium on Biomineralization*, P. Westbroek and E. de Jong, eds. Reidel Publishing Co., Dordrecht, Holland. Pp. 15–25.

Lovelock, J.E. 1988. *The Ages of Gaia.* W.W. Norton Co., New York.

Lovelock, J.E. and L. Margulis. 1974. Atmospheric homeostasis, by and for the biosphere: The Gaia hypothesis. *Tellus* **26**:1–10.

Lovelock, J.E. and A.J. Watson. 1982. The regulation of carbon dioxide and climate: Gaia or geochemistry. *Journal of Planetary Science* **30**:795–802.

Lowenstam, H.A. 1974. Impact of life on chemical and physical processes. In: *The Sea.* Vol. 5, *Marine Chemistry*, E. Goldberg, ed. J. Wiley, New York. Pp. 715–796.

Lowenstam, H.A. and S. Weiner. 1989. *On Biomineralization.* Oxford University Press, New York.

Mah, R.A., D.M. Ward, and L. Baresi. 1977. Biogenesis of methane. *Annual Review of Microbiology* **31**:309–342.

Margulis, L. 1969. New phylogenies of the lower organisms: Possible relation to organic deposits in Precambrian sediment. *Journal of Geology* **77**:606–617.

Margulis, L. 1982. *Early Life.* Science Books International, Boston.

Margulis, L. and J.E. Lovelock. 1974. Biological modulation of the Earth's atmosphere. *Icarus* **21**:471–489.

Margulis, L. and J.F. Stolz. 1983. Microbial systematics and a Gaian view of the sediments. In: *The Fourth International Symposium on Biomineralization*, P. Westbroek and E. de Jong, eds. Reidel Publishing Co., Dordrecht, Holland. Pp. 27–54.

Margulis, L., L. Lopez Baluja, S.M. Awramik, and D. Sagan. 1986. Community living long before man: Fossil and living microbial mats and early life. In: *Man's Effect on the Global Environment*, D. Botkin and A.A. Orio, eds. Vol. 1. Elsevier Science Publishers, Amsterdam.

Maynard-Smith, J. 1982. *Evolution Now a Century after Darwin.* Macmillan Press, London.

McCrae, W.H. 1981. Long time-scale fluctuation in the evolution of the Earth. *Proceedings of the Royal Society* Ser. A, **375**:1–41.

Napier, W.M. and S.V.M. Clube. 1979. A theory of terrestrial catastrophism. *Nature* **282**:455–459.

Redfield, A.C. 1958. The biological control of chemical factors in the environment. *American Scientist* **46**:205–221.

Riggs, D.S. 1970. *Control Theory and Physical Feedback Mechanisms.* Williams and Wilkins, Baltimore.

Sagan, D. and L. Margulis. 1983. The Gaian perspective of ecology. *The Ecologist: Journal of the Post Industrial Age* **13**:160–167.

Sagan, D. 1986a. Towards a global metabolism—the sulphur cycle. *The Ecologist: Journal of the Post Industrial Age* **16**:14–17.

Sagan, D. 1986b. Sulfur: Toward a global metabolism. *The Science Teacher* **53**:15–20.

Sagan, D., ed. 1985. *The Global Sulfur Cycle.* NASA Technical Memorandum 87570. Washington, DC 20546.

Schopf, J.W., ed. 1983. *Earth's Earliest Biosphere.* Princeton University Press, Princeton, New Jersey.

Shukla, J. and Y. Mintz. 1982. Influence of the land-surface evapotranspiration on the Earth's climate. *Science* **215**:1498–1501.

Sillen, L.-G. 1966. Regulation of O_2, N_2 and CO_2 in the atmosphere: Thoughts of a laboratory chemist. *Tellus* **18**:198–206.

Synge, R.L.M. 1983. Personal communication.

Vernadsky, V.I. 1929. *La biosphere*. Alcan, Paris.

Vernadsky, V.I. 1986. *The Biosphere* (an abridged version based on the French edition of 1929). Synergetic Press.

Walker, J.C.G., P.B. Hays and J.F. Keating. 1981. A negative feedback mechanism for the long-term stabilization of the Earth's surface temperature. *Journal of Geophysical Research* **86**:9776–9782.

Waddington, C.H. 1976. Concluding remarks in *Evolution and Consciousness*, E. Jantsch and C.H. Waddington, eds. AddisonWesley Publishing Co., Reading, Massachusetts. Pp. 243–249.

Watson, A. and J.E. Lovelock. 1983. Biological homeostasis of the global environment: The parable of "daisy world." *Tellus* **35B**:284–289.

Watson, A., J.E. Lovelock, and L. Margulis. 1978. Methanogenesis, fires and the regulation of atmospheric oxygen. *BioSystems* **10**:293–298.

Westbroek, P. and E. de Jong, eds. 1983. *Biomineralization and Biological Mineral Precipitation*. Reidel Publishing Co., Dordrecht, Holland. Pp. 27–54.

Woese, C. and G. Fox. 1977. The concept of cellular evolution. *Journal of Molecular Evolution* **10**:1–6.

II   THE INTEGRAL BIOSPHERE

JOHN F. STOLZ, DANIEL B. BOTKIN, AND
MINOO N. DASTOOR

PARTS OF THE BIOSPHERE

The biosphere is a remarkable planetary life support system. Extending from the bottom of the oceans to the upper limits of the troposphere, it is a large-scale system of integrated parts that contains and sustains life. The purpose of this chapter is to describe these parts and to introduce the reader to the ways they interact to form this functioning life support system.

The biosphere can be most simply divided into four parts; the hydrosphere, the lithosphere, the atmosphere, and the biota. The hydrosphere is the watery portion of the Earth that includes the oceans and seas. The lithosphere is the solid Earth, the continents, islands, and atolls. The atmosphere is that gaseous envelope around the surface of the globe. The sum total of all life on the Earth forms the biota. Only the portions of the hydrosphere, lithosphere, and atmosphere in direct contact with the biota are part of the biosphere; they are called the environment (Botkin et al., 1984).

The biological components of the biosphere, the biota, may be categorized in several ways: 1) by genealogy and evolutionary relationships, 2) by the external form and internal structure, and 3) on the basis of their functions or activities in the biosphere. These functions include the particular chemical reactions that a kind of life carries out and the actions that lead either to removal or to addition of a chemical compound to a nonliving portion of the biosphere. These functions also include the ways that one kind of life can influence the function of another.

From the viewpoint of genealogy and evolution, life is made up of three basic types: 1) viruses, 2) prokaryotes (bacteria), and 3) eukaryotes (everything else). Viruses are not made up of cells, contain only a few proteins and genetic material (DNA or RNA), and are obligately parasitic on cellular life forms. The cellular forms of life (prokaryotes and eukaryotes) have been further classified into five kingdoms (Table II-1, Margulis and Schwartz, 1982).

Prokaryotes are microorganisms that do not have a membrane-bounded nucleus but rather contain their DNA (the genetic material) in the general cellular material (Figure II-1). Prokaryotes divide by binary fission, multiple fission, or budding. They do not have organelles, such as mitochondria. The important point about prokaryotes from a planetary perspective is that they carry out a diversity of chemical reactions and have a great variety of metabolic capabilities. Biochemical pathways that are exclusively prokaryotic include the production of methane (methanogenesis), nitrogen fixation, and sulfate reduction. Autotrophy, the production of organic compounds from inorganic ones, occurs in prokaryotes. Some representatives are photosynthetic, using energy derived from sunlight. Chemosynthetic bacteria derive energy from inorganic compounds. Heterotrophy, the use of organic compounds as a carbon source, and mixotrophy, where cellular carbon is derived from both carbon dioxide and organic compounds, are also modes of metabolism in bacteria. Prokaryotes are the sole occupants of the Moneran kingdom.

Eukaryotes are those organisms whose cells have the bulk of their genetic material (DNA) in a membrane-bounded nucleus and have intracellular organelles, such as mitochondria, chloroplasts, lysosomes, and endoplasmic reticulum (Figure II-1). Reproduction may occur asexually, usually by mitosis (but also sometimes by budding and binary fission), or sexually. Eukaryotes also exhibit a wide range of metabolic capabilities including photosynthesis, heterotrophy, and mixotrophy. Eukaryotic organisms have been divided into four kingdoms based on phylogeny (evolutionary relatedness), metabolism, and reproduction. These are the Protoctista, Fungi, Plantae, and Animalia (Margulis and Schwartz, 1982).

The kingdom Protoctista includes algae, protozoa, flagellated fungi, and slime molds. There is a great diversity of metabolic processes (i.e., photoauto-, hetero-, and mixotrophy) and reproductive strategies (i.e., binary fission, budding, mitosis, meiosis) within the protists.

TABLE II-1   THE CLASSIFICATION OF ORGANISMS

Acellular life: Viruses—RNS or DNA containing (nonautopoietic)[a]

Cellular life (autopoietic):

Prokaryotes	Eukaryotes
cell wall, nucleoid, flagella, 70s ribosomes, binary or multiple fission	membrane-bounded nucleus, organelles (i.e., golgi), mitochondria, chloroplasts, 80s ribosomes, mitosis, meiosis

KINGDOM MONERA:
  Prokaryotic cells, nutrition—absorptive, chemosynthetic, photoheterotrophic or photoautotrophic. Metabolism—anaerobic, facultative, microaerophilic or aerobic. Motile and nonmotile forms. Flagella composed of flagellin. Bacteria. Cyanobacteria.

KINGDOM PROTOCTISTA:
  Eukaryotic cells with solitary and colonial unicellular organization. Nutrition—ingestive, absorptive or photoautotrophic (with chloroplasts). Premitotic and eumitotic asexual reproduction, sexual reproduction. Undulipodia and cilia with tubulin microtubules ("9 + 2"). Lack embryos and complex cell junctions. Nucleated algae, protozoans, chytrids, hypochytrids, myxomycetes, oomycetes, slime molds, slime nets.

KINGDOM FUNGI:
  Eukaryotic, primarily multinucleate, coenocytic, septate, and nonseptate. Asexual and sexual reproduction; zygotic meiosis. Nutrition absorptive, no motility. Zygomycetes, Ascomycetes, Basidiomycetes, Deuteromycetes.

KINGDOM ANIMALIA:
  Multicellular eukaryotes, develop from diploid blastula, genetic meiosis, nutrition heterotrophic, digestive phagocytosis. Differentiation. Metazoa, etc.

KINGDOM PLANTAE:
  Eukaryotic, O_2 producing photoautotrophy, develop from embryos, tissue differentiation, alternation of diploid and haploid generation, chloroplasts (plastids). Bryophytes, angiosperms, gymnosperms, etc.

[a] For the concept of autopoiesis, see Fleischaker, G. and L. Margulis, 1987. *Autopoiesis and the Origin of Life*. COSPAR Proceedings 1986, Toulouse, France. Advances in Space Science (in press).

The Fungi kingdom is made up of eukaryotic organisms that have mycelial development and includes the true fungi (e.g., mushrooms, bread mold). Fungi are heterotrophic and saprophytic (absorb degrading organic material), and aid in the decomposition of organic materials and nutrient cycling in forest litters.

The Plant kingdom is comprised of mosses, liverworts, and the vascular plants (e.g., angiosperms and gymnosperms). Plants are primarily photoautotrophic, reproduce sexually, and develop from an embryo. From a planetary perspective, plants are important in the cycling of water and ele-

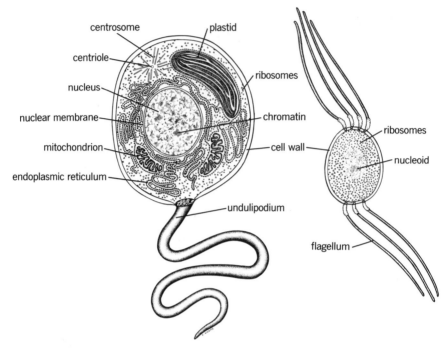

centrosome   plastid

centriole

nucleus

ribosomes

nuclear membrane

chromatin

mitochondrion

ribosomes

cell wall

nucleoid

endoplasmic reticulum

undulipodium

flagellum

FIGURE II–1   Comparison between prokaryote and eukaryote cell (based on drawing from Margulis, 1981).

ments such as carbon, nitrogen, potassium, phosphorus, and sulfur. They are also important for their role in soil stabilization and for their effects on climate. Plants remove atmospheric carbon dioxide and convert it into usable carbon compounds. In this process they also produce oxygen. Trees and higher plants are instrumental in the transpiration of water and their root systems prevent soil erosion. These characteristics, plus the effects of large stands of vegetation (i.e., forests) on the reflectivity of the Earth's surface (albedo) all contribute to global climate.

Members of the Animal kingdom are heterotrophic and ingest their food. They reproduce sexually, although budding and parthenogenesis (self-fertilization) may occur. Animals are important in food webs and biomineralization.

The members of the biota may also be categorized by function. Most simply, there are producers of organic compounds, consumers, and decomposers. Primary producers are those organisms that derive energy from either sunlight or inorganic compounds and make organic compounds from carbon dioxide. Many primary producers like plants, algae, and photosynthetic bacteria, obtain the energy for making organic compounds from sunlight (photosynthesis). Others use the energy stored in the chemical bonds of inorganic compounds (chemoautotrophy), such as sulfur-oxidating bacteria. Consumers cannot make organic compounds from inorganic ones, but can transform some organic compounds into others. They obtain cellular carbon and energy from these transformations. Consumers include animals, protozoa, and many bacteria. Decomposers live on dead organisms and transform organic compounds back into inorganic ones, including carbon dioxide and water. Fungi and some bacteria are common decomposers. Together, producers, consumers, and decomposers and their local environment function as spatial units with more or less loosely defined boundaries. These units are called ecosystems.

ECOSYSTEM

The smallest unit of the biosphere that has all the characteristics to sustain life is the ecosystem. An ecosystem may be defined as an assemblage of populations grouped into communities and interacting with each other and their local environment. A salt marsh with its characteristic vegetation (i.e., *Spartina*, *Salicornia*), animal and microbial life, and nutrient cycles (as occur in food webs), is an example of an ecosystem. Communities are interacting populations of individuals belonging to different species. Microbial mat communities, for instance, are composed of a great variety of microorganisms with different metabolic capabilities. Populations are made up of organisms that are members of the same species, occurring at the same time and in the same place.

Ecosystems come in a large variety of sizes, and the smallest (e.g., microbial mats) may be only a few centimeters whereas the largest (e.g., tundra) may be measured in kilometers. These boundaries are sensitive to climatic and other variables, and are critical points for the exchange of information and energy between ecosystems. No natural ecosystem is entirely independent; ions and particulates in suspension, propagules, and other matter transfer between them.

Ecosystems are characterized by a diversity of species. There must be representatives from the three functional or metabolic groups: primary producers, consumers, and decomposers. In any ecosystem, certain species are more important than others in terms of sustaining the necessary flow of energy and materials for the likely persistence of other life forms. Ecologists have long speculated that the diversity of species within an ecosystem insures the long-term survival of that ecosystem because the diversity itself provides a redundancy of function. The mass or abundance of a species may be misleading as to its overall function in the ecosystem. Organisms may have effects that are disproportionate to their relative biomass.

Life is a patchy phenomenon, and the patches may themselves have importance for life's persistence. This patchiness exists at many size scales so that we can think of life as forming patches of varying density with variable intensity. At the microscale, small gradients play a large role. For example, physically stratified microbial communities establish gradients of oxygen, pH, redox potential (Eh), and light on the millimeter scale (Cohen et al., 1984). Interfaces are also essential. For example, the transition zone between the aerobic and anaerobic zones is often the location of significant biogeochemical cycling (Krumbein, 1983). Other interfaces of importance are sediment-water and water-air.

Ecological communities exhibit temporal heterogeneity involving changes at time intervals that range from minutes to thousands of years. Organisms with short generation times can respond to environmental changes rapidly. A sudden influx of organics may trigger a bloom of microorganisms within hours. At the other extreme, organisms with long generation times and life spans may continue to persist long after conditions optimal for them have gone. Ecosystems, then, are not static; they are dynamic systems in time and space. They should not be viewed as steady state systems, but rather as systems that continue to evolve biologically and change environmentally.

ECOSYSTEM TYPES

What kind of organism lives in a particular place depends on the environment and on certain aspects of the history of our planet, which has determined what organisms have been able to travel where. There are physical and metabolic adaptations to each kind of Earth environment, and this leads to another way to categorize the biota, that is, by external shape and internal form. This is called physiognomic classification. With this in mind, we can define a biome as a physiognomic class of a set of ecosystems. A biome describes a set of ecosystems within a geographical region exposed to the same climatic conditions and having dominant species with a similar life cycle, climatic adaptations, and physical structure (Botkin et al., 1986).

An example of a biome is the tropical savannah. Tropical savannahs are composed of grasslands, plus a mixture of trees and shrubs in a semiarid

TABLE II-2 CHARACTERISTIC PRIMARY PRODUCERS OF
TERRESTRIAL AND AQUATIC BIOMES

TERRESTRIAL	AQUATIC
Tropical rain forest	Open oceans
Tropical seasonal forest	Upwelling zones
Temperate evergreen forest	Continental shelf
Temperate deciduous forest	Algal bed
Borreal forest	Coral reef
Woodland and shrubland	Thermal vents*
Savanna	Estuaries
Temperate grassland	Swamp and marsh
Tundra and alpine meadow	Lake and rivers
Desert scrub	Littoral marine
Rock, ice, and sand	Continental shelf or slope benthic*
Cultivated land	Abyssal benthic*
Urban	

*Below photic zone, therefore, chemoautotrophic primary producers.

climate (Odum, 1971). In the biosphere there are terrestrial and aquatic biomes.

Terrestrial biomes have often been defined by the vegetation types that dominate the community (physiognomy) (Table II–2). The types of vegetation affect the climate and soil structure, and characterize the particular biome. The tropical rain forests of Central and South America, for example, receive substantial amounts of rain throughout the year (80 to 90 inches), and the soil has a low organic content. Nutrients are quickly cycled, and there is a rapid loss of fertility when the endogenous vegetation is removed. These forests are dominated by broad-leaved evergreens, tree ferns, lianas, epiphytic orchids, and bromeliads (Whittaker, 1970). Terrestrial vegetation has a rapid exchange with the atmosphere of oxygen, water, and carbon dioxide. The concentration of carbon dioxide in the atmosphere is affected by terrestrial vegetation seasonally and annually.

Aquatic ecosystems fall into two categories: fresh water and marine. Lakes, rivers, marshes, and swamps are considered part of continental biomes. Marine biomes include open ocean, continental shelf, deep-sea vent and upwelling zones. Estuaries are unique in that they lie at the interface of the terrestrial and marine biomes. Hypersaline ecosystems are unique as well (Friedman and Krumbein, 1985). They are dominated by a complex microbiota, many species of which require exceedingly high concentrations of salt.

PROPERTIES OF THE BIOSPHERE

The biosphere has three basic properties: 1) it harnesses the necessary energy to run the system, 2) it maintains the supply of life's essential elements, and 3) it is capable of changing in response to cosmological, geological, and biological perturbations. In this section, each of these properties will be discussed in turn.

Energy is needed to maintain structure and biomass in the biosphere. The flow of energy is accomplished through the interaction of primary producers, consumers, and decomposers. Most of this energy is derived from sunlight, while a small portion of it is derived from inorganic processes. The primary producers harness this energy and provide the rest of the biota with reduced carbon. Primary productivity is limited to the Plant Kingdom and certain groups of protoctists and bacteria (Table II–3). Trees, higher plants, and algae (i.e., the phototrophic protoctists) take sunlight, carbon dioxide, and water and produce carbohydrates and oxygen. In the Kingdom Monera, there is oxygenic photosynthesis, as well as a type of photosynthesis where oxygen is not involved (i.e., anoxic photosynthesis). In addition, certain bacteria can derive chemical energy through the oxidation of inorganic compounds in the absence of light and use carbon dioxide as the source of carbon (chemoautotrophy).

The type of primary producer found in an ecosystem depends on the environment. Plants, for example, have three different ways of making carbon compounds from carbon dioxide. Depending on the carbon compound they initially produce, plants can be categorized as either C_3 (phosphoglycerate, a three-carbon compound), C_4 (oxaloacetate, a four-carbon compound), or CAM (cressulacean acid metabolism). These three pathways are adaptations by vegetation to different environments. C_3 plants have a relatively high rate of photorespiration. They can lose 20 percent to 50 percent of their fixed carbon during photorespiration and are associated with high altitudes. C_4 plants lose less CO_2 via photorespiration and are usually found in hot, dry climates and at intermediate altitudes. CAM plants are usually found in desert communities at low altitudes and employ a diurnal cycle of CO_2 fixation. Because stomatal opening occurs at night, less water is lost during CO_2 fixation, providing CAM plants with the drought resistance necessary for surviving in an arid environment.

Aquatic environments are the realm of few vascular plants. Instead, algae, the phototrophic members of the Protoctista kingdom, are the predominant primary producers. Although they have been categorized into as many as twelve different types, these algae fall into three basic groups: 1) the red algae (rhodophytes), 2) the green algae (chlorophytes), and 3) the brown algae (phaeophytes). Traditionally, the larger members of the pho-

TABLE II-3 PRIMARY PRODUCERS[a]: EXAMPLES OF MAJOR GROUPS

PLANT KINGDOM[c]
C3 plants (maple), C4 plants (maize), CAM plants (succulents)

PROTOCTISTA KINGDOM[d]
Rhodophyta (red algae), Chlorophyta (green algae), Gamophyta (conjugating green algae), Euglenophyta (euglenoids), Phaeophyta (brown alga), Eustigmatophyta (eustigonemes), Bacillariophyta (diatoms), Chrysophyta (golden algae), Pyrophyta (dinoflagellates), Xanthophyta (golden brown algae), Haptophyta (coccoliths), Cryptophyta (cryptomonads)

MONERAN KINGDOM[e]
Chloroxybacteria *(Prochloron)*, Cyanobacteria *(Oscillatoria)*, Chromatiacea[b] *(Chromatium)*, Rhodospirillaceae[b] *(Rhodospirillum)*, Chlorobiaceae[b] *(Chlorobium, Chloroflexus)*; Chemoautotrophs *(Thiobacillus)*

[a] Compiled from Margulis and Schwartz, 1982.
[b] Denotes anoxic photosynthesis.
[c] Over 500,000 species estimates grouped into over 300 families in 10 phyla
[d] Phyla of the protoctista kingdom (Margulis, et al. 1989)
[e] Families and genera of the Moneran (Prokaryotae) Kingdom, all subkingdom Eubacteria, four phyla represented

totrophic protoctista have been called seaweeds (i.e., kelp), while the smaller, microscopic algae have been called phytoplankton or, if even smaller, nannoplankton (Odum, 1971). Much of the primary productivity of the oceans is attributed to such algae.

The phototrophic bacteria may be separated grossly into two groups, those that produce oxygen and those that do not (Starr et al., 1981). They are a diverse group of organisms that utilize an assortment of pigments and light-capturing devices to harness the sun's energy. They differ both morphologically and biochemically, depending on whether they produce oxygen in the process of photosynthesis. In fact, oxygen is toxic to those phototrophic bacteria that do not produce it. The phototrophic bacteria are found everywhere as films and scums on the surface of lakes and ponds. They may be most prominent in environments that restrict the growth of higher vegetation and grazing by predators (e.g., metazoans). Examples of such environments are hot springs, evaporite flats, arctic dry valleys, and certain intertidal zones and salt marshes (Cohen et al., 1984).

Not all primary productivity is photosynthetic in nature. The discovery of thermal vent communities at the East Pacific Rise and similar systems along the base of the Florida Escarpment shows that chemoautotrophic bacteria alone can provide the necessary energy and carbon to an ecosystem (Desbryeres and Lambier, 1983; Jannash and Wilson, 1979; Paull et al., 1984).

The continued availability of biologically important elements is necessary for the long-term survival of life. Carbon, hydrogen, nitrogen, oxygen, sulfur, and phosphorus are the elements with which the nucleic acids (RNA and DNA), amino acids (proteins), carbohydrates, and lipids found in the cell are made. There are more than a dozen elements needed in trace amounts (i.e., Ca^{++}, Mg^{++}, Fe^{++}, Na^+, K^+, Cu^{++}, Co^+, Mo^+, Ni^+, Cl^-). The availability of an element depends on the geological and biological effects on the chemical state of the element, namely, its roles within biogeochemical cycles. Biogeochemical cycles are important not only because life depends on them, but also because life affects the biosphere greatly through the flow and transformation of chemical compounds (Moore and Dastoor, 1984).

It is difficult to imagine the cycling of a particular element as simply circular. An element may go through its various redox states via a circuitous route, and there are the inevitable interactions with other elemental cycles. From a thermodynamic view, the cycle begins with the element in its highest energy state (usually reduced), followed by a series of oxidations and ending in its most oxidized state with a resulting increase in entropy. An input of energy is necessary in order to return the element to the original reduced state. In reality, it is a lot more complicated. Take the nitrogen cycle, for example (Figure II–2). In the environment, nitrogen exists as (from the most reduced state to the most oxidized) NH_4, N_2, N_2O, NO_2, and NO_3. Biological transformation of nitrogen occurs via three processes: 1) nitrogen fixation (N_2 to NH_4), 2) nitrification (NH_4 to NO_2 to NO_3), and 3) nitrate reduction (NO_3 to NO_2 to N_2O to N_2).

The carbon cycle is important for providing assimilable carbon to the biosphere. Several of the gaseous phases are also instrumental in the modulation of climate (i.e., methane and carbon dioxide are greenhouse gases). Carbon is fixed as CO_2 by the activity of photo- and chemoautotrophic organisms. This reduced carbon is transformed back to CO_2 by aerobic respiration or anaerobic respiration and fermentation. Methanogenesis and methylotrophic oxidation of methane convert CO_2 to methane and methane back to CO_2.

The biological transformation of sulfur occurs by two processes: sulfate reduction (SO_4 to H_2S) and sulfur oxidation (H_2S to S' to S'' to S_2O_3 to SO_3 to SO_4). The phosphorus cycle is complicated by the fact that on the Earth phosphorus is never found in a gaseous state. Therefore, it commonly exists in particulate form as phosphate cycled in and out of solution. Table II–4 is a summary of the prokaryotic roles in the biogeochemical cycles of carbon, nitrogen, and sulfur.

Life affects the composition of sediments through the precipitation of

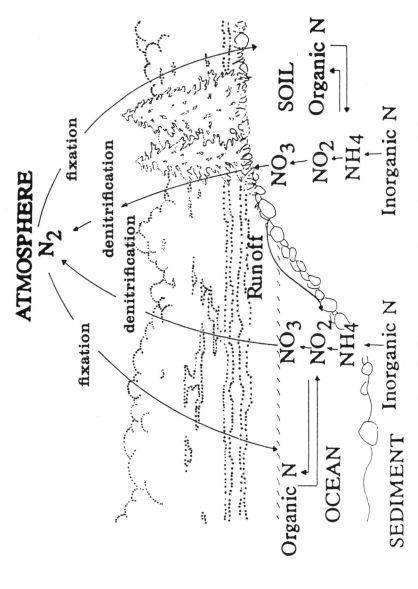

FIGURE II–2   The global nitrogen cycle (based on Blackburn, 1983).

TABLE II-4 BIOGEOCHEMICAL CYCLES IN PROKARYOTES

Element	Process	Examples of organisms and summary of reactions
CARBON:	CO_2 fixation:	$CO_2 + H_2 \rightarrow (CH_2O)n + A_2$ (A = O, S)
		photoautotrophs: cyanobacteria, purple and green sulfur bacteria
		chemoautotrophs: sulfur and iron oxidizing bacteria
	Methanogenesis:	$COO^- + H_2 \rightarrow CH_4$ methanogenic bacteria
	Methylotrophy:	$CH_4 + O_2 \rightarrow CO_2$ methylotrophic bacteria
	Fermentation:	$(CH_2O)n \rightarrow CO_2$ anaerobic heterotrophic bacteria
	Respiration:	$(CH_2O)n \ O_2 \rightarrow CO_2$ aerobic heterotrophic bacteria
SULFUR:	Sulfur reduction:	$SO_4 = + H_2 \rightarrow H_2S$ sulfur reducing bacteria
	Sulfur oxidation:	$H_2S \rightarrow S^o$ purple and green sulfur phototrophs
		$S^o + O_2 \rightarrow SO_4$ sulfur oxidizing bacteria
NITROGEN:	N_2 fixation:	$N_2 + H_2 \rightarrow NH_4$ phototrophic bacteria, nitrogen-fixing heterotrophic bacteria
	Nitrification:	$NH_4 + O_2 \rightarrow NO_2, NO_3$ nitrifying bacteria
	Denitrification:	$NO_2, NO_3 \rightarrow N_2O, N_2$ denitrifying heterotrophic bacteria

minerals (Margulis and Stolz, 1983). Some forty different minerals are now known to be precipitated by organisms (Table II–5, based on Lowenstam and Weiner, 1983; see also Table AI-2, pp. 152-156). Biomineralization is the term used to describe the processes by which minerals are deposited by organisms. Minerals may be precipitated indirectly, as a result of metabolic processes (i.e., *biologically induced*), or directly, under conditions determined by the organisms and under genetic control (i.e., *matrix mediated*). Biologically induced biomineralization occurs when the conditions for precipitation of a mineral phase occur as a result of biological activity. For example, in some sedimentary environments the hydrogen sulfide produced by bacteria in the sediment may react with iron oxides resulting in the formation of iron sulfides (i.e., pyrite) (Krumbein, 1983). Matrix-mediated biomineralization usually occurs when the composition, shape, and size of the mineral is essential to the function. The magnetosome of bacteria that can respond to changes in the Earth's magnetic field is a remarkable example. The bacteria take iron in solution and precipitate crystals of magnetite (Fe_3O_4) of uniform size and shape, and align them into a chain (Blakemore, 1982). The chain of magnetite crystals has the properties of a perfect bar magnet. Bone is another example of this type of biomineralization.

The rates by which a mineral is precipitated may be accelerated by biological activity. For years oceanographers pondered how manganese nodules formed on the surface of the ocean floors. It had been calculated that the manganese nodules were being deposited more slowly than the surrounding ocean floor sediments, but the nodules were not being buried. The discovery of manganese oxidizing bacteria provided a plausible solution. These bacteria catalyze the precipitation of manganese oxides (i.e., birnesite) enzymatically or by producing organic compounds that enhance autooxidation of manganese both within and outside the cells (Nealson, 1983).

Biogenic minerals may be deposited under physical and chemical conditions that preclude the precipitation of the mineral or mineral phase in the absence of life. An example of this is the mussel shell. In the environment that the mussel lives, the calcium carbonate used to form the shell should be calcite and not aragonite, which it is. As another example of matrix-mediated biomineralization, the shell-forming organ of the mussel, the nacreous layer, not only deposits calcium carbonate, but also determines how the crystals will be aligned (Weiner and Traub, 1980).

The effects of biomineralization can be dramatic. Large deposits of limestone are the result of the precipitation of calcium carbonate by coccolithophorids and forams, phototrophic protoctists that produce calcium carbonate scales (Westbroek and de Jong, 1983). The removal of carbon dioxide from the atmosphere and the concurrent burial of the carbon affects the composition of the atmosphere, oceans, and sediment as well as climate. Overall, the concentrations of the dissolved trace metals, inorganic phosphorus, silica, nitrate, and carbonate in seawater and ocean basins are known to be strongly affected by the marine biota (Whitfield and Watson, 1983). For example, the concentration of silica in ocean waters has been greatly decreased by the precipitation of silicious spicules in sponges and silicious frustules in diatoms.

TABLE II-5 THE DIVERSITY AND DISTRIBUTION OF BIOGENIC MINERALS THROUGHOUT THE FIVE KINGDOMS

	PROKARYOTAE (MONERA)	PROTOCTISTA	FUNGI	PLANTAE	ANIMALIA
CARBONATES	calcite aragonite monohydrocalcite	calcite aragonite vaterite a.h. carbonate[1]		calcite aragonite vaterite	calcite aragonite vaterite monohydrocalcite protodolomite a.h. carbonate[1]
PHOSPHATES	dahllite	dahllite	dahllite	dahllite?	francolite dahllite $Ca_3Mg_3(PO_4)_4$ huntite brushite octacalcium phosphate calcium pyrophosphate a. dahllite prec.[2] a. brushite prec.[3] a. whitlockite prec.[4] a. Fe-Ca-phosphate[5]
HALIDES					fluorite a. fluorite prec.[6]
OXALATES		whewellite weddelite	whewellite weddelite glushinskite	whewellite weddelite	whewellite weddelite
CITRATES					calcium citrate

44

SULFATES		gypsum, celestite, barite	gypsum	gypsum	gypsum
SILICA		opal	opal	opal	opal
Fe-OXIDES	magnetite, lepidocrocite? ferrihydrite, amor. ferrihydrite[7]	magnetite?, amor. ferrihydrite[7]	ferrihydrite	ferrihydrite	magnetite, geothite, lepidocrocite, ferrihydrite, amor. ferrihydrite[7], amor. ilmenite[8]
Mn-OXIDES	todorokite, birnessite				
SULFIDES	pyrite, hydrotroilite, sphalerite, wurtzite, galena				

The term *precursor* refers to an amorphous phase that, upon heating to 500° C, converted to the designated crystalline form.

[1] Amorphous hydrous carbonate
[2] Amorphous dahllite precursor
[3] Amorphous brushite precursor
[4] Amorphous whitlockite precursor
[5] Amorphous Fe-Ca phosphate
[6] Amorphous fluorite precursor
[7] Amorphous ferrihydrite
[8] Amorphous ilmenite

Organisms may affect the dissolution of minerals as well as their precipitation. Lichens are known to affect the weathering rates of rocks by secreting organic acids, which break down the surface. Endolithic organisms and their predators (grazing snails, limpets, and chitons) have a dramatic effect on the weathering of carbonate rocks. Boring algae and bacteria will colonize the surface of a rock, penetrating the top few millimeters. Then, grazing organisms, like the chiton with its hard, sharp teeth, will remove a millimeter or so of the surface, digest the algae, and excrete the carbonate rock (Golubic, 1973).

The chemical properties of an element determine the way organisms affect it. For example, iron is soluble under anoxic conditions, but becomes insoluble when oxidized. Microbial activity in sediments produce anoxic zones, which reduce oxidized iron minerals and render the iron soluble. In many environments, the presence of both organisms capable of oxidizing and organisms capable of reducing a particular element insures the continued availability of the element. Organic coatings produced by organisms may also enable minerals to persist under otherwise unfavorable conditions. Particulates formed on the surface of the ocean contain reduced minerals even in fully oxygenated waters (Buat-Menard and Che-selet, 1979).

The present biosphere has been shaped by cosmological, geological, and biological phenomena over the course of the history of the Earth. The Earth has changed over the past 4.5 billion years from a planet whose surface was constantly bombarded with planetesimals to one with continents and plate tectonics. There has been an increase in the luminosity of the sun and large meteorites still occasionally strike the surface. The biota has not only responded to these changes, but has also brought about some dramatic changes in the biosphere through biological innovations. For example, the evolution of oxygenic photosynthesis has had a profound effect on the biota and on the chemistry of the Earth's surface. That the biota has survived in the face of change is a manifestation of the collective strength of the biosphere; its ability to respond to changes at all periodicities and amplitudes has been one way that life has sustained itself.

The thread of life has continued uninterrupted from about 3.8 billion years ago to the present. It is marked by certain events that had a dramatic effect on the biota and the biosphere. The initial one was the origin of life itself. The first evidence for life, the microfossils from South Africa and Australia, are about 3.5 billion years old (Awramik, 1984). These microfossils are the remains of bacteria.

Bacteria dominated the biota for the first two billion years. Many of the biological pathways necessary for the survival of the biota evolved during that time. The first of these was the evolution of photosynthesis. This insured that the biota would never be short of a carbon source. In a world without atmospheric oxygen, the first photosynthetic organisms were anaerobic. The evolution of oxygen-producing photosynthesis caused a dramatic change in the biosphere. The oxygen produced was toxic to most organisms. Also, many of the essential elements are insoluble when oxidized.

New environments were created and colonized by organisms that could adapt. Some organisms were able to adapt better when in close association with other organisms. The first eukaryotic organisms appeared about 1.5 billion years ago. The fossilized remains of planktonic algae have been found in sediments of that age. Eukaryotic cells evolved, most probably through a series of symbiotic associations (Margulis, 1981). The mitochondria and chloroplasts of eukaryotic cells were probably once free-living bacteria. They are more closely related structurally and biochemically to bacteria than to the rest of the cell.

Once eukaryotic cells had evolved, it was possible for larger, multicellular organisms to evolve. The colonization of land by animals and plants expanded the boundaries of the biosphere. Today, the presence of life is evident virtually everywhere it has been looked for on the surface of the Earth, from the Antarctic dry valleys to the Sahara Desert to the Galapagos vents. The present-day biota has a long and uninterrupted history, and has been shaped by its past. It continues to evolve in and with its environment.

SUMMARY

The biosphere is the life support system of the planet. To understand how it works, we have identified its components. The basic unit of the biosphere, the ecosystem, has certain characteristics. These characteristics have to be studied not only on the organismal and systems level, but also from a historical and planetary perspective. At the organismal level, the biota was shown to be made up of two kinds of cellular life. The simpler forms gave rise to more complex forms through interaction and symbiosis. As a system, the biosphere functions to maintain structure through the input of energy and to insure the continued availability of life's essential elements. Historically, the biosphere has functioned to sustain life since life first evolved some three and a half billion years ago. From the planetary perspective, this planet is the only one in the inner solar system that has a biosphere and can sustain life. Establishing a science of the biosphere will bring us to understand the biosphere better and hopefully insure our place in it in the future.

REFERENCES

Awramik, S. 1984. Ancient stromatolites and microbial mats. In: *Microbial mats: stromatolites*, Y. Cohen, R.W. Castenholz, H.O. Halvorson, ed. Alan R. Liss Inc., New York, pp. 1–22.

Berthelin, J. 1983. Microbial weathering processes. In: *Microbial Geochemistry*, W.E. Krumbein, ed. Blackwell Scientific Publications, Boston, pp. 223–262.

Blackburn, H.T. 1983. The microbial nitrogen cycle. In: *Microbial Geochemistry*, W.E. Krumbein, ed. Blackwell Scientific Publications, Boston, pp. 63–90.

Blakemore, R. 1982. Magnetotactic bacteria. *Annual Reviews Microbiology* **36**:217–238.

Botkin, D.B., M. Davis, J. Estes, A. Knoll, R.V. O'Neill, L. Orgel, L.B. Slobotkin, J.C.G. Walker, J. Walsh, and D.C. White. 1986. *Remote Sensing of the Biosphere*, National Academy Press, Washington, DC, 135 pp.

Botkin, D., J.E. Estes, R.M. MacDonald, and M.V. Wilson. 1984. Studying the Earth's vegetation from space. *BioScience* **34**:508–514.

Buat-Menard, P. and P. Chesselet. 1979. Variable influence of the atmospheric-flux on the trace metal chemistry on ocean suspended matter. *Earth and Planetary Letters* **42**:399–411.

Cohen, Y., R.W. Castenholz, and H.O. Halvorson. 1984. *Microbial Mats: Stromatolites*. Alan R. Liss Inc., New York.

Desbryeres, D. and L. Laubier. 1983. Primary consumers from hydrothermal vents and animal communities. In: *Hydrothermal Processes at Seafloor Spreading Centers*, P.A. Rona, K. Bostroem, L. Laubier, and K.L. Smith, eds. Plenum Press, New York, pp. 711–734.

Friedman, G.M. and W.E. Krumbein. 1985. *Hypersaline Ecosystems: The Gavish Sabkha*. Springer-Verlag, Berlin.

Gest, H. and J.L. Favinger. 1983. Heliobacterium chlorum, an anoxygen brownish-green photosynthetic bacterium containing a "new" form of bacteriochlorophyll. *Archives of Microbiology* **136**:11–16.

Golubic, S. 1973. The relationship between blue-green algae and carbonate deposits. In: *The Biology of Blue-Green Algae*, N.G. Carr and B.A. Whitton, eds. Blackwell Scientific Publications, Boston, pp. 434–472.

Jannash, H.W. and C.O. Wirsen. 1979. Chemosynthetic primary production at East Pacific sea-floor spreading center. *Bioscience* **29**:592–598.

Krumbein, W.E., ed. 1983. *Microbial Geochemistry*. Blackwell Scientific Publications, Boston.

Lowenstam, H. and S. Weiner. 1983. Mineralization by organisms and the evolution of biomineralization. In: *Biomineralization and Biological Metal Accumulation*, P. Westbroek and E.W. de Jong, eds. Reidel Publishing Co., The Netherlands, pp. 191–204.

Margulis, L. 1981. *Symbiosis and Cell Evolution*. W.H. Freeman and Co., San Francisco.

Margulis, L. and K.V. Schwartz. 1982. *The Five Kingdoms: Guide to the Phyla of Life on Earth*. W.H. Freeman and Co., San Francisco.

Margulis, L. and J.F. Stolz. Microbial systematics and a Gaian view of the sediments. In: *Biomineralization and Biological Metal Accumulation*, P. Westbroek and E.W. de Jong, eds. Reidel Publishing Co., The Netherlands, pp. 15–53.

Margulis, L., J.O. Corliss, M. Melkonian, and D. Chapman, eds. 1989. *Handbook of Protoctista. The structure, cultivation, habitats and life cycles of the eukaryotic microorganisms and their descendants, exclusive of animals, plants and fungi. A guide to the algae, ciliates, foraminifera, sporozoa, water molds, slime molds and other protoctists*. Jones and Bartlett Publishers, Inc, Boston.

Moore, B. and M.N. Dastoor, eds. 1984. *The Interaction of Global Biochemical Cycles*. NASA JPL Publication 84–21.

Nealson, K.N. 1983. The microbial manganese cycle. In: *Microbial Geochemistry*, W.E. Krumbein, ed. Blackwell Scientific Publications, Boston, pp. 191–222.

Odum, E. 1971. *Fundamentals of Ecology*. W.B. Saunders Co., Philadelphia.

Paull, C.K., B. Hecker, R. Commeau, R.P. Freeman-Lynde, C. Neumann, W.P. Corso, S. Golubic, J.E. Hook, E. Sikes, and J. Curry. 1984. Biological communities at the Florida Escarpment resemble hydrothermal vent taxa. *Science* **226**:965–967.

Starr, M.P., H. Stolp, H.G. Trueper, A. Balows, and H.G. Schlegel. 1981. *The Prokaryotes*. Springer-Verlag, Berlin.

Watson, A.J., J.E. Lovelock, and L. Margulis. 1978. Methanogenesis, fires and the regulation of atmospheric oxygen. *BioSystems* **10**:293–298.

Weiner, S. and Traub, W. 1980. X-ray diffraction study of the insoluble organic matrix of mollusc shells. *FEBS Letters* **111**:311–316.

Westbroek, P. and W.E. de Jong, eds. 1983. *Biomineralization and Biological Metal Accumulation*. Reidel Publishers, The Netherlands.

Whitfield, M. and A.J. Watson. 1983. The influence of biomineralization on the composition of seawater. In: *Biomineralization and Biological Metal Accumulation*. Reidel Publishers, The Netherlands, pp. 57–72.

Whittaker, R.H. 1970. *Communities and Ecosystems*. The Macmillan Co., New York.

III PHOTOCHEMISTRY OF BIOGENIC GASES

JOEL S. LEVINE

THE ATMOSPHERE

The Earth's atmosphere extends to several thousand kilometers above the surface, where it eventually merges with interplanetary space. The total mass of the atmosphere (5.1×10^{21} grams) is very small compared to the total mass of the oceans (1.39×10^{24} grams) and the total mass of the Earth (5.98×10^{27} grams) (Walker, 1977). Yet this thin gaseous envelope that surrounds our planet has been, and is today, of critical importance to life. Beginning about 4.6 billion years ago with the process of chemical evolution (i.e., the synthesis of complex organic molecules, the precursors of living systems, from the simple gases in the prebiological paleoatmosphere), we see the beginning of the continuing interaction between life and the atmosphere.

Once life formed, it was protected from biologically lethal solar ultraviolet radiation by the presence of atmospheric ozone, which is produced via photochemical processes from oxygen, a biogenic gas (Levine, 1985). Over the history of the Earth, living organisms have significantly modified and, to this day, continue to modify the chemical composition of the atmosphere (Lovelock and Margulis, 1974; Margulis and Lovelock, 1974, 1978; Lovelock, 1979). Margulis and Lovelock (1978) have diagrammed the biogenic gases produced metabolically by various microorganisms. Their list

51

includes the following gases: nitrogen (N_2), oxygen (O_2), carbon dioxide (CO_2), nitrous oxide (N_2O), ammonia (NH_3), methane (CH_4), carbon monoxide (CO), hydrogen (H_2), hydrogen sulfide (H_2S), and dimethyl sulfide [$(CH_3)_2S$]. To this list the following biogenic gases may be added: nitric oxide (NO), dimethyl disulfide [$(CH_3)_2S_2$], and the methyl halogens (i.e., methyl chloride (CH_3Cl), methyl bromide (CH_3Br), and methyl iodide (CH_3I)). For some of these gases, biogenic production is the overwhelming source (e.g., oxygen, nitrous oxide, ammonia, methane, hydrogen sulfide, dimethyl sulfide, and dimethyl disulfide). For others, the strength of the biogenic source is not accurately known, but is probably significant (e.g., carbon monoxide, nitric oxide, hydrogen, and the methyl halogens). All these biogenic species, even at trace levels, impact the photochemistry and chemistry of the lower and upper atmosphere (Graedel, 1985; Turco, 1985); lead to the formation of acid precipitation and atmospheric aerosols (Graedel, 1985); and affect the climate of the Earth via the greenhouse effect (Kuhn, 1985). The composition of the atmosphere is summarized in Table III–1, with the major source or sources for each gas (i.e., biogenic, volcanic, photochemical, or anthropogenic) given in the third column (Levine, 1985).

The atmosphere can be divided into an electrically neutral region (consisting of neutral atoms and molecules), which accounts for almost the entire mass of the atmosphere, and a region of electrically charged particles (consisting of electrons and charged atoms and molecules) superimposed on the neutral atmosphere. The neutral atmosphere is further subdivided into five main regions: the troposphere, the stratosphere, the mesosphere, the thermosphere, and the exosphere. These regions are defined by their temperature gradient, as shown in Figure III–1. The exosphere, not shown on Figure III–1, is an isothermal region that begins at about 500 km and eventually merges with interplanetary space. Superimposed on the structure of the neutral atmosphere are several regions of high concentrations of electrons and charged atoms and molecules resulting from the ionization of neutral gases by high energy solar (X-rays and ultraviolet) radiation and cosmic rays. The regions of charged particles are found in four layers collectively termed the ionosphere: the D-layer (below about 90 km), the E-layer (90 km–120 km), the F-1 layer (a daytime feature centered at about 150 km), and the F-2 layer (centered at about 300 km). The magnetosphere, or radiation belts, another region of charged particles mostly of solar origin and contained by the Earth's magnetic field, extends out to several planetary radii. With two notable exceptions (i.e., water vapor and ozone), the gross composition of the atmosphere is fairly constant up to about 100 km, although the number density of the atmosphere (molecules per cm^3) decreases exponentially with altitude.

TABLE III-1 THE COMPOSITION OF THE TROPOSPHERE

	Concentration[a]	Source[b]
A. *Major and Minor Gases*		
Nitrogen (N_2)	78.08%	Volcanic, biogenic
Oxygen (O_2)	20.95%	Biogenic
Argon (Ar)	0.93%	Radiogenic
Water vapor (H_2O)	Variable—up to 4%	Volcanic, evaporation
Carbon dioxide (CO_2)	0.034%	Volcanic, biogenic, anthropogenic
B. *Trace Gases*		
1. *Oxygen Species*		
Ozone (O_3)	10–100 ppbv	Photochemical
Atomic oxygen (O) (ground state)	10^3 cm$-^3$	Photochemical
Atomic oxygen [$O(^1D)$] (excited state)	10^{-2} cm$-^3$	Photochemical
2. *Hydrogen Species*		
Hydrogen (H_2)	0.5 ppmv	Photochemical, biogenic
Hydrogen peroxide (H_2O_2)	10^9 cm^{-3}	Photochemical
Hydroperoxyl radical (HO_2)	10^8 cm^3	Photochemical
Hydroxyl radical (OH)	10^6 cm^{-3}	Photochemical
Atomic hydrogen (H)	1 cm^{-3}	Photochemical
3. *Nitrogen Species*		
Nitrous oxide (N_2O)	330 ppbv	Biogenic, anthropogenic
Ammonia (NH_3)	0.1–1 ppbv	Biogenic, anthropogenic
Nitric acid (HNO_3)	50–1000 pptv	Photochemical
Hydrogen cyanide (HCN)	⁓200 pptv	Anthropogenic(?)
Nitrogen dioxide (NO_2)	10–300 pptv	Photochemical
Nitric oxide (NO)	5–100 pptv	Anthropogenic, biogenic, lightning, and photochemical
Nitrogen trioxide (NO_3)	100 pptv	Photochemical

TABLE III-1   THE COMPOSITION OF THE TROPOSPHERE (*continued*)

	Concentration[a]	Source[b]
PAN ($CH_3CO_3NO_2$)	50 pptv	Photochemical
Dinitrogen pentoxide (N_2O_5)	1 pptv	Photochemical
Pernitric acid (HO_2NO_2)	0.5 pptv	Photochemical
Nitrous acid (HNO_2)	0.1 pptv	Photochemical
Nitrogen aerosols:		
Ammonium nitrate (NH_4NO_3)	⁓10 pptv	Photochemical
Ammonium chloride (NH_4Cl)	⁓0.1 pptv	Photochemical
Ammonium sulfate (($NH_4)_2SO_4$)	⁓0.1 pptv(?)	Photochemical
4. Carbon Species		
Methane (CH_4)	1.7 ppmv	Biogenic, anthropogenic
Carbon monoxide (CO)	70–200 ppbv (N. Hemisphere) 40–60 ppbv (S. Hemisphere)	Anthropogenic, biogenic, photochemical
Formaldehyde (H_2CO)	0.1 ppbv	Photochemical
Methylhydroperoxyl radical (CH_3OOH)	10^{11} cm^{-3}	Photochemical
Methylperoxyl radical (CH_3O_2)	10^8 cm^{-3}	Photochemical
Methyl radical (CH_3)	10^{-1} cm^{-3}	Photochemical
5. Sulfur Species		
Carbonyl sulfide (COS)	0.5 ppbv	Volcanic, anthropogenic
Dimethyl sulfide (($CH_3)_2S$)	0.4 ppbv	Biogenic
Hydrogen sulfide (H_2S)	0.2 ppbv	Biogenic, anthropogenic
Sulfur dioxide (SO_2)	0.2 ppbv	Volcanic, anthropogenic, photochemical
Dimethyl disulfide (($CH_3)_2S_2$)	100 pptv	Biogenic
Carbon disulfide (CS_2)	50 pptv	Volcanic, anthropogenic

TABLE III-1 THE COMPOSITION OF THE TROPOSPHERE (*continued*)

	Concentration[a]	Source[b]
Sulfuric acid (H_2SO_4)	20 pptv	Photochemical
Sulfurous acid (H_2SO_3)	20 pptv	Photochemical
Sulfoxyl radical (SO)	10^3 cm^{-3}	Photochemical
Thiohydroxyl radical (HS)	1 cm^{-3}	Photochemical
Sulfur trioxide (SO_3)	10^{-2} cm^{-3}	Photochemical
6. *Halogen Species*		
Hydrogen chloride (HCl)	1 ppbv	Sea salt, volcanic
Methyl chloride (CH_3Cl)	0.5 ppbv	Biogenic, anthropogenic
Methyl bromide (CH_3Br)	10 pptv	Biogenic, anthropogenic
Methyl iodide (CH_3I)	1 pptv	Biogenic, anthropogenic
7. *Noble Gases (chemically inert)*		
Neon (Ne)	18 ppmv	Volcanic (?)
Helium (He)	5.2 ppmv	Radiogenic
Krypton (Kr)	1 ppmv	Radiogenic
Xenon (Xe)	90 ppbv	Radiogenic

[a] Species concentrations are given in percent by volume or in terms of surface mixing ratio—parts per million by volume (ppmv = 10^{-6}), parts per billion by volume (ppbv = 10^{-9}), parts per trillion by volume (pptv = 10^{-12}), or in terms of surface number density (cm^{-3}). The species mixing ratio is defined as the ratio of the number density of the species to the total atmospheric number density (2.55×10^{19} molecules cm^{-3}). There is some uncertainty in the concentrations of species at the ppbv level or less. The species concentrations given in molecules cm^{-3} are generally based on photochemical calculations and species concentrations in mixing ratios are generally based on measurements.

[b] Species major sources are divided into the broad category of volcanic, radiogenic, biogenic, anthropogenic, and photochemical.

About 80 percent to 85 percent of the total mass of the atmosphere resides in the troposphere. The troposphere is in direct contact with the biosphere and, hence, regulates or modulates the transfer of gases and particulates into and out of the biosphere. Almost all of the water vapor in the atmosphere is found in the troposphere, where its distribution is controlled by the evaporation-condensation cycle. The troposphere is the

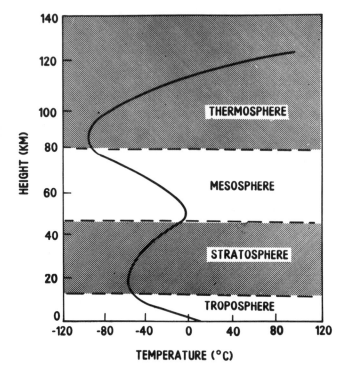

FIGURE III-1   The structure of the atmosphere. Each atmospheric region is defined by its temperature gradient.

region of weather phenomena (i.e., cloud formation and precipitation). Precipitation is a source of dissolved water-soluble gases and particulates (i.e., nitrites, nitrates, sulfates, chlorides, and ammonium) for the biosphere. Most of the remaining mass of the atmosphere is located in the stratosphere, directly above the troposphere. About 90 percent of the ozone in the atmosphere is located in the stratosphere. The absorption of biologically lethal solar ultraviolet radiation (200 nm–300 nm) by ozone makes the stratosphere of great importance to the biosphere and to life.

This chapter will deal with the relationship between the biosphere and the atmosphere and, hence, will concentrate on the composition and photochemistry and chemistry of the troposphere and stratosphere, which, combined, represent more than 99 percent of the total mass of the Earth's atmosphere. It is within the troposphere, and to a lesser extent within the stratosphere, that biogenic gases produced at the Earth's surface are photochemically and chemically transformed to other species and are eventually returned to the biosphere to complete the biogeochemical cycle.

This chapter will review the photochemical and chemical processes in the troposphere and stratosphere that transform biogenic gases and recycle them back to the biosphere. Further information on the photochemical processes and parameters and the kinetic reaction rates for the photochemical and chemical reactions discussed in this chapter may be found in Graedel (1979; 1980; 1985), Logan et al. (1981), Levine and Allario (1982), and Turco (1985).

OXYGEN, OZONE, AND HYDROXYL

The most important biogenic species in the atmosphere is molecular oxygen (O_2) produced as a by-product during the process of photosynthesis, which can be expressed by the following reaction:

$$nH_2O + mCO_2 + h\nu \xrightarrow{\text{chlorophyll}} C_m(H_2O)_n + mO_2 \quad (1)$$

In equation (1), $C_m(H_2O)_n$ represents carbohydrate produced by the green plant cell from water vapor (H_2O) and carbon dioxide (CO_2) in the presence of sunlight, represented by $h\nu$, where h is Planck's constant and ν is the frequency of the visible solar radiation. The carbohydrate produced in the photosynthetic process is utilized as food by the plant.

There is general agreement that photosynthetic activity and the subsequent biogenic production of O_2 was responsible for transforming the oxidation state of the early atmosphere from a mildly reducing atmosphere of carbon dioxide, nitrogen (N_2), and water vapor to the present, strongly oxidizing mixture of nitrogen and oxygen (for example, see Walker, 1977; Levine, 1982, 1985). However, some O_2 could have been produced abiotically in the prebiological paleoatmosphere due to the photolysis of H_2O and CO_2, the two major components of volcanic outgassing (Walker, 1978; Levine et al., 1982; Canuto et al., 1982; Levine, 1985). The photochemical reactions leading to the abiotic production of O_2 can be expressed as[2]:

$$H_2O + h\nu \rightarrow OH + H; \lambda \leq 240 \text{ nm} \quad (2)$$

followed by:

$$OH + OH \rightarrow O + H_2O \quad (3)$$

and

$$CO_2 + h\nu \rightarrow O + CO; \lambda \leq 230 \text{ nm} \quad (4)$$

---

[2] The wavelength corresponds to the photodissociation threshold for the reaction, or the minimum energy required for the reaction to occur.

The atomic oxygen (O) produced in reactions (3) and (4) forms O_2 via the following reactions:

$$O + O + M \rightarrow O_2 + M, \tag{5}$$

and

$$O + OH \rightarrow O_2 + H \tag{6}$$

where OH is the hydroxyl radical, CO is carbon monoxide, H is atomic hydrogen, and M is any molecule to absorb excess energy and/or momentum of the reaction. The abiotic production of O_2 via reactions (2) to (6) is very sensitive to atmospheric levels of H_2O and CO_2, and to the flux of solar ultraviolet radiation, all of which may have varied significantly over geological time (Canuto et al., 1982). The vertical distribution of O_2 in the prebiological paleoatmosphere, for present atmospheric levels of H_2O, CO_2, and solar ultraviolet radiation, is shown in Figure III–2 (Levine et al., 1982). For present values of these parameters, the surface mixing ratio of O_2 is about 10^{-15}.[3] Note that in the prebiological paleoatmosphere, O_2 was not uniformly distributed with altitude in the troposphere and stratosphere, as it is in the present atmosphere, but varied significantly with altitude. The altitude distribution of O_2 reflects its chemical production and destruction terms. For enhanced levels of CO_2 and solar ultraviolet radiation, the surface mixing ratio of O_2 may have approached the parts per billion by volume level (ppbv = 10^{-9}) (Canuto et al., 1982). However, it is still significantly below the present atmospheric level of about 21 percent by volume. It should be pointed out that, in addition to the production reactions shown in equations (2) to (6), the calculations shown in Figure III–2 also include various O_2 loss reactions (i.e., the photolysis of O_2 and losses due to various photochemical and chemical processes).

In addition to transforming the early atmosphere from a mildly reducing mixture to a highly oxidizing one, the biogenic growth of O_2 in the atmosphere had another important biological effect—it led to the photochemical production of ozone (O_3), which eventually resulted in the shielding of the Earth's surface from biologically lethal solar ultraviolet radiation. The photochemical production of O_3 is initiated by the photolysis of O_2-form-

---

[3]Trace species concentrations are usually given in terms of surface mixing ratio — parts per million by volume (ppmv = 10^{-6}), parts per billion by volume (ppbv = 10^{-9}), or parts per trillion by volume (pptv = 10^{-12}). The surface species mixing ratio is defined as the ratio of the surface number density of the species (atoms or molecules cm^{-3}) to the total surface number density of the atmosphere (2.55×10^{19} molecules cm^{-3}).

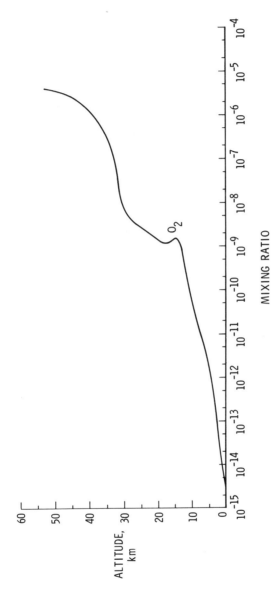

FIGURE III–2   The vertical distribution of O_2 in the prebiological paleoatmosphere (Levine et al., 1982). Prior to the origin and evolution of photosynthetic organisms, oxygen resulted from the photolysis of volcanic H_2O and CO_2 by solar ultraviolet radiation.

ing oxygen atoms (O), which then may recombine to form O_3. These reactions can be expressed as:

$$O_2 + h\nu \rightarrow O + O; \lambda \leq 242 \text{ nm} \tag{7}$$

followed by:

$$O + O_2 + M \rightarrow O_3 + M \tag{8}$$

The calculated vertical distribution of O_3 as a function of atmospheric O_2 level, expressed in present atmospheric level (PAL) of O_2, is shown in Figure III–3 (Levine, 1982). For biological shielding of solar ultraviolet, the critical parameter is the total atmospheric burden or column density (O_3 molecules cm^{-2}) of O_3, as opposed to the vertical profile of O_3 shown in Figure III–3. The total O_3 column density as a function of atmospheric O_2 level expressed in PAL is given in Table III–2 (Levine, 1982). Biological shielding is obtained for a total O_3 column density of about 6×10^{18} O_3 molecules cm^{-2} (Berkner and Marshall, 1965), which is approximately half of the present total O_3 column density of about 10^{19} O_3 molecules cm^{-2}, which varies with latitude and season. The calculations given in Figure III–2 and Table III–2 include the various O_3 loss reactions (i.e., the photolysis of O_3 and the chemical losses due to reactions with oxides of nitrogen, hydrogen, and chlorine, as well as the production reaction given in equation (8)).

In the present atmosphere, O_3 not only shields the surface of the Earth from biologically lethal solar ultraviolet radiation, but also initiates a photochemical reaction that leads to the chemical transformation of almost every biogenically produced gas. Biogenic gases are transformed via oxidation by the hydroxyl radical (OH). The hydroxyl radical is produced by the reaction of excited oxygen [O(^{1}D)] with water vapor. Excited oxygen results from the photolysis of O_3. These processes can be represented by the following reactions:

$$O_3 + h\nu \rightarrow O(^1D) + O_2; \lambda \leq 310 \text{ nm} \tag{9}$$

$$O(^1D) + H_2O \rightarrow 2 \text{ OH} \tag{10}$$

The transfer of solar radiation through the atmosphere that leads to the production of O(^{1}D) via the photolysis of O_3 (reaction (9)) is controlled by molecular absorption, multiple scattering (due to atmospheric molecules and aerosol particles), and surface albedo. Photochemical calculations of the vertical distribution of O(^{1}D) in the present troposphere are shown in Figure III–4 (Augustsson and Levine, 1982). The vertical distribution of OH in the present troposphere, initated by reaction (10), is shown in Fig-

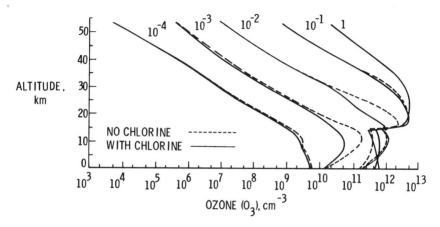

FIGURE III-3 The vertical distribution of O_3 as a function of the level of atmospheric O_2, represented in terms of present atmospheric level (PAL) (Levine, 1982). Photosynthetic activity was the overwhelming source of O_2 and was responsible from transforming the atmosphere from a reducing to oxidizing one.

ure III-5 (Augustsson and Levine, 1982). Both sets of calculations were performed for different values of surface albedo, ranging from 0 percent (total absorption by the surface) to 100 percent (total reflection by the surface). The global albedo of the Earth is about 25 percent to 30 percent. It should be pointed out that O_3 may also be photolyzed by solar radiation

TABLE III-2 OZONE IN THE EARLY AND PRESENT ATMOSPHERE

O_2 Level (PAL)	O_3 Column Density (cm^{-2})	Height of O_3 Peak (km)	O_3 Density at Peak (cm^{-3})
1. *Without chlorine species chemistry*			
1	9.93 (18)[a]	20.5	5.53 (12)
10^{-1}	6.07 (18)	19	4.57 (12)
10^{-1}	2.47 (18)	16	2.48 (12)
10^{-3}	1.88 (17)	11.5	1.92 (11)
10^{-4}	5.58 (15)	0	5.63 (09)
2. *With chlorine species chemistry*			
1	9.70 (18)	20.5	5.40 (12)
10^{-1}	5.94 (18)	19	4.62 (12)
10^{-2}	1.59 (18)	10	1.16 (12)
10^{-3}	6.98 (16)	9	5.72 (10)
10^{-4}	5.18 (15)	0	5.42 (09)

[a] 9.93 (18) is read as 9.93×10^{18}.

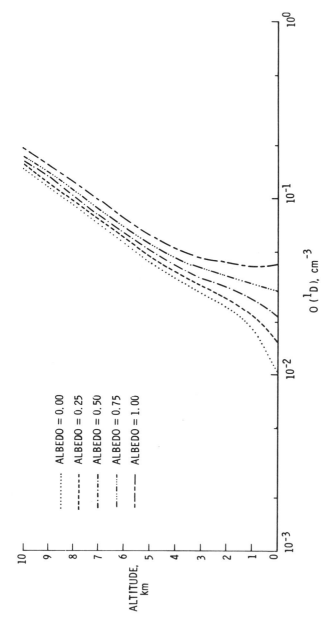

FIGURE III-4 The vertical distribution of excited oxygen, O(^{1}D) in the present troposphere. These theoretical calculations include the effects of molecular absorption, multiple scattering (due to molecules and aerosols), and surface albedo. Calculations were performed for different values of surface albedo, ranging from 0 to 100% (Augustsson and Levine, 1982).

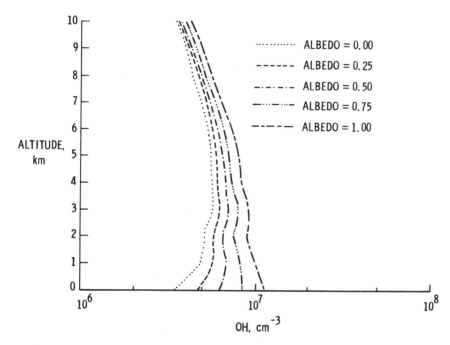

FIGURE III–5 The vertical distribution of the hydroxyl radical (OH) in the present troposphere. Calculations are similar to those in Figure III–4.

of wavelength greater than 310 nm, but this reaction leads to the production of ground state atomic oxygen (O) rather than the more energetic excited oxygen [$O(^1D)$]. Reaction (10), leading to the formation of OH, is the overwhelming OH production term in the present troposphere, since solar photons between about 290 nm–310 nm can reach the Earth's surface (stratospheric O_3 absorbs solar photons less than about 290 nm and, hence, they cannot reach the troposphere). However, in the O_2 and O_3 deficient paleoatmosphere, solar photons less than 290 nm could easily reach the surface and the direct photolysis of H_2O (reaction (2)) was a major source of OH in the prebiological paleoatmosphere.

The reaction of OH with various biogenic species initiates reactions that lead to the chemical transformation of these biogenic gases. The reaction of biogenic gases with OH represents the major chemical destruction mechanism for these biogenic species. In the following section, we will consider the details of the photochemistry and chemistry of the biogenic species initiated by the reaction with OH and discuss the fate of these biogenic gases in the atmosphere.

THE FATE OF BIOGENIC GASES IN THE ATMOSPHERE

AMMONIA (NH$_3$)

After molecular nitrogen (N$_2$) and nitrous oxide (N$_2$O: mixing ratio = about 330 ppbv), ammonia is the most abundant nitrogen species with a highly variable surface mixing ratio (due to its short atmospheric residence time of about ten days and localized sources), ranging from less than 0.1 to about 10 ppbv (10^{-9}), depending on season and meteorological and soil conditions (Hoell et al., 1982). Ammonia is the only gaseous basic constituent of the atmosphere, a usually acidic environment due to the presence of the following acids: carbonic (H$_2$CO$_3$), nitric (HNO$_3$), nitrous (HNO$_2$), and sulfuric (H$_2$SO$_4$). By virtue of its high solubility, NH$_3$ neutralizes the ever-present tropospheric acids and, hence, controls the pH of cloud droplets and the acidity of rain and snow. In addition to its role in the chemistry of the atmosphere to be discussed later in this section, NH$_3$ absorbs thermal infrared radiation around 10.53 μm and, hence, affects the climate of the Earth (Wang et al., 1976).

Thermodynamic equilibrium calculations indicate that the atmospheric equilibrium mixing ratio of NH$_3$ should be about 10^{-35} (Lovelock and Margulis, 1974). The measured atmospheric mixing ratio of about 10^{-9}, a 26-order-of-magnitude enhancement over the theoretical equilibrium mixing ratio, is a striking testimony to the role of microorganisms in maintaining drastic disequilibrium conditions in the composition of the atmosphere. In addition to biogenic production, the overwhelming source of NH$_3$, other nonbiogenic sources include coal conversion and combustion processes, and certain industrial processes and activities.

The chemical destruction of NH$_3$ is controlled by its reaction with OH:

$$NH_3 + OH \rightarrow NH_2 + H_2O \tag{11}$$

This reaction leads to the formation of the amine radical (NH$_2$). Subsequent reactions of the amine radical may lead to either the production or destruction of the oxides of nitrogen (NO$_x$ = NO + NO$_2$) via the following reactions (Logan et al., 1981):

$$NH_2 + O_3 \rightarrow NO_x + products \tag{12}$$

$$NH_2 + NO \rightarrow N_2 + H_2O \tag{13}$$

$$NH_2 + NO_2 \rightarrow N_2O + H_2O \tag{14}$$

For NO$_x$ concentrations below about 60 pptv, reactions (12) to (14) lead to a net production of NO_x, whereas, for NO_x concentrations greater than 60 pptv, these reactions lead to a net destruction of NO_x (Logan et al., 1981).

As already mentioned, the reaction of NH_3 with OH (reaction (11)) is the major chemical mechanism for tropospheric NH_3 with a characteristic chemical destruction time of about 40 days for a surface OH density of about 10^6 cm^{-3} (Levine et al., 1980). However, the atmospheric loss of NH_3 is controlled by rainout with a characteristic loss time of about ten days (Levine et al., 1980). Another heterogeneous process that leads to the atmospheric loss of NH_3 is the formation of ammonium nitrate (NH_4NO_3) and ammonium sulfate [$(NH_4)_2SO_4$] aerosols via the following reactions:

$$NH_3(g) + HNO_3(g) \rightarrow NH_4NO_3(s) \tag{15}$$

and

$$2NH_3(g) + H_2SO_4(g) \rightarrow (NH_4)_2SO_4(s) \tag{16}$$

(g denotes gaseous phase; s denotes solid phase).

The loss of atmospheric NH_3 via rainout, aerosol formation, and dry deposition is a source of ammonium ions (NH_4^+) to the biosphere.

Due to its rapid loss due to rainout and aerosol formation and reaction with OH (reaction (11)), NH_3 decreases rapidly through the troposphere and very little, compared to surface levels, reaches the stratosphere (Hoell et al., 1982).

METHANE (CH_4)

After carbon dioxide (CO_2: mixing ratio = 330 ppmv), methane, with a mean global mixing ratio of about 1.66 ppmv, is the most abundant carbon species. Due to its relatively long atmospheric residence time (about eight years), CH_4 appears to be uniformly mixed with respect to altitude within the troposphere, but exhibits a slight latitudinal gradient (about 1.72 ppmv north of 20°N and about 1.62 ppmv south of the intertropical convergence zone). In situ measurements (Graedel and McRae, 1980; Rasmussen and Khalil, 1981) since the late 1970s, analyses of historic ground-based solar infrared spectra providing information on time scales of decades (Rinsland et al., 1985), and chemical analyses of gases trapped in ice cores providing information, although somewhat uncertain, on time scales of hundreds to thousands of years (Craig and Chou, 1982), all indicate that tropospheric levels of methane have increased significantly, with the current annual increase about 1 percent to 2 percent per year.

Methane absorbs thermal infrared radiation at about 7.7 μm (Wang et al., 1976). An increase in CH_4 from 0.7 ppmv to its present value of about 1.66 ppmv may have caused an increase in the global temperature of the Earth of about 0.23° C (Wang et al., 1976), which is about half of the temperature increase calculated to have occurred as a result of increases in atmospheric CO_2.

Thermodynamic equilibrium calculations indicate that the equilibrium mixing ratio of CH_4 should be about 10^{-35} (Lovelock and Margulis, 1974). The measured atmospheric mixing ratio of about 10^{-6} is a 29-order-of-magnitude enhancement due to the production of CH_4 by biogenic processes. Methane is produced by fermentation of organic matter in anoxic environments, such as swamps, tropical rain forests, and rice paddies (Harriss and Sebacher, 1981). Other sources of CH_4 include enteric fermentation in ruminants (mostly cattle), biomass burning, natural gas leakage (Ehhalt and Schmidt, 1978), and termites (Zimmerman et al., 1982).

Reaction with OH is the overwhelming loss mechanism for CH_4 via the following reaction:

$$CH_4 + OH \rightarrow CH_3 + H_2O \qquad (17)$$

This reaction leads to the formation of the methyl radical (CH_3). Subsequent reactions of the methyl radical, initiated by reaction (17), form the methane oxidation scheme, which is summarized below (for example, see Logan et al., 1981):

$$CH_3 + O_2 + M \rightarrow CH_3O_2 + M \qquad (18)$$
$$(CH_3O_2 = \text{methylperoxyl radical})$$

$$CH_3O_2 + NO \rightarrow CH_3O + NO_2 \qquad (19)$$
$$(CH_3O = \text{methoxyl radical})$$

$$NO_2 + h\nu \rightarrow NO + O; \lambda \le 400 \text{ nm} \qquad (20)$$

$$O + O_2 + M \rightarrow O_3 + M \qquad (8)$$

$$CH_3 + O_2 \rightarrow H_2CO + HO_2 \qquad (21)$$
$$(H_2CO = \text{formaldehyde}; HO_2 = \text{hydroperoxyl radical})$$

$$HO_2 + NO \rightarrow NO_2 + OH \qquad (22)$$

$$NO_2 + h\nu \rightarrow NO + O; \lambda \le 400 \text{ nm} \qquad (20)$$

$$O + O_2 + M \rightarrow O_3 + M \qquad (8)$$

$$H_2CO + h\nu \rightarrow CO + H_2, \text{ or } HCO + H; \lambda \le 340 \text{ nm} \qquad (23)$$
$$(HCO = \text{formyl radical})$$

The net cycle for the methane oxidation scheme can be represented as:

$$CH_4 + 4O_2 \rightarrow H_2O + CO + H_2 + 2O_3 \text{ (net cycle)} \qquad (24)$$

REDUCED SULFUR SPECIES

Hydrogen sulfide (H_2S), methyl sulfide [$(CH_3)_2S$], and methyl disulfide [$(CH_3)_2S_2$] are biogenically produced sulfur species with surface mixing ratios of about several tenths part per billion by volume. These gases are produced by bacteria in anaerobic, sulfate-rich environments, such as marine sediments and coastal mud flats. Reaction with OH is the major chemical loss for these gases. The oxidation of H_2S by OH leads to the production of sulfur dioxide (SO_2) and eventually sulfuric acid (H_2SO_4), the major constituent of acid rain. The reaction scheme of $(CH_3)_2S$ and $(CH_3)_2S_2$ with OH is less certain, but is also believed to lead to the formation of SO_2 and H_2SO_4. The reactions describing the oxidation of H_2S to H_2SO_4 are summarized here (Graedel, 1979):

$$H_2S + OH \rightarrow HS + H_2O \qquad (25)$$
(HS = thiohydroxyl radical)

$$HS + O_2 \rightarrow SO + OH \qquad (26)$$
(SO = sulfoxyl radical)

$$SO + O_2 \rightarrow SO_2 + O \qquad (27)$$

$$SO_2 + OH + M \rightarrow HSO_3 + M \qquad (28)$$
(HSO_3 = sulfuric acid radical)

$$HSO_3 + OH \rightarrow H_2SO_4 \qquad (29)$$

Several other reaction schemes have been suggested for the transformation of SO_2 to H_2SO_4. These include:

$$SO_2 + HO_2 \rightarrow SO_3 + OH \qquad (30)$$
(SO_3 = sulfur trioxide)

$$SO_3 + H_2O \rightarrow H_2SO_4 \qquad (31)$$

REDUCED HALOGEN SPECIES

It has been suggested that methyl chloride (CH_3Cl: 0.5 ppbv), methyl bromide (CH_3Br: 10 pptv), and methyl iodide (CH_3I: 1 pptv) are produced biogenically in the ocean as well as by anthropogenic activities. A mechanism for the oxidation of CH_3Cl by OH is described here (Graedel, 1979):

$$CH_3Cl + OH \rightarrow CH_2Cl + H_2O \tag{32}$$
$$(CH_2Cl = \text{chloromethyl radical})$$

$$CH_2Cl + O_2 + M \rightarrow CH_2ClO_2 + M \tag{33}$$

$$CH_2ClO_2 + NO \rightarrow CH_2ClO + NO_2 \tag{34}$$

$$CH_2ClO + O_2 \rightarrow HCl + CO + HO_2 \tag{35}$$
$$(HCl = \text{hydrogen chloride})$$

Similar reaction schemes are probably initiated by the oxidation of CH_3Br and CH_3I by OH.

CARBON MONOXIDE (CO)

The mixing ratio of carbon monoxide (CO), which exhibits a strong latitudinal gradient (70 ppbv–200 ppbv in the Northern Hemisphere and 40 ppbv–60 ppbv in the Southern Hemisphere), suggests a strong, if not dominant, anthropogenic source (Logan et al., 1981). The magnitude of the biogenic source of CO remains uncertain. Atmospheric CO has an atmospheric residence time of about three months and is controlled by its reaction time with OH, which initiates the carbon monoxide oxidation scheme, summarized here (for example, see Logan et al., 1981):

$$CO + OH \rightarrow CO_2 + H \tag{36}$$

$$H + O_2 + M \rightarrow HO_2 + M \tag{37}$$

$$HO_2 + NO \rightarrow NO_2 + OH \tag{38}$$

$$NO_2 + h\nu \rightarrow NO + O; \lambda \leq 400 \text{ nm} \tag{20}$$

$$O + O_2 + M \rightarrow O_3 + M \tag{8}$$

Assuming sufficient levels of NO for reaction (38) to proceed, the net cycle for the carbon monoxide oxidation scheme can be represented as:

$$CO + 2O_2 \rightarrow CO_2 + O_3 \text{ (net cycle)} \tag{39}$$

Continuous in situ measurements (Khalil and Rasmussen, 1984) and analysis of historic ground based solar infared spectra (Rinsland and Levine, 1985) indicate that atmospheric levels of CO may be increasing between about 1 percent to 5 percent per year. This CO increase combined with increasing levels of CH_4 suggest that tropospheric levels of OH may have decreased by about 25 percent over the last 35 years (Levine et al., 1985).

NITRIC OXIDE (NO)

Nitric oxide (NO) is biogenically produced by several bacteria (Lipschultz et al., 1981; Anderson and Levine, 1986, 1987). The magnitude of the biogenic source is uncertain and may be comparable to (Galbally and Roy, 1978) or significantly less than the anthropogenic source of NO (i.e., high temperature combustion processes are believed to be the major source of NO (Crutzen, 1979)). Nitric oxide is rapidly converted to nitrogen dioxide (NO_2) by reacting with O_3 and HO_2:

$$NO + O_3 \rightarrow NO_2 + O_2 \qquad (40)$$

$$NO + HO_2 \rightarrow NO_2 + OH \qquad (41)$$

The photolysis of NO_2 by visible solar radiation in the troposphere leads to the rapid formation of NO:

$$NO_2 + h\nu \rightarrow NO + O; \lambda \leq 400 \text{ nm} \qquad (20)$$

The reaction of NO with OH leads to the formation of nitrous acid (HNO_2):

$$NO + OH + M \rightarrow HNO_2 + M \qquad (42)$$

The NO_2 produced in reactions (40) and (41) reacts with OH to form nitric acid (HNO_3), the fastest increasing component of acid rain:

$$NO_2 + OH + M \rightarrow HNO_3 + M \qquad (43)$$

The loss of water-soluble HNO_3, which has a characteristic atmospheric residence time of about three days, is controlled by rainout. The loss of HNO_3 and HNO_2 by rainout and other heterogeneous loss mechanisms (e.g., dry deposition and aerosol formation) are the major loss mechanisms for the oxides of nitrogen ($NO_x = NO + NO_2$) in the atmosphere. Nitric and nitrous acids are sources of fixed nitrogen (i.e., nitrates (NO_3^-) and nitrites (NO_2^-)) to the biosphere. Once in the biosphere, NO_3^-, NO_2^-, and NH_4^+ are recycled into the atmosphere in the forms of molecular nitrogen (N_2), nitrous oxide (N_2O), and NO via nitrification and denitrification.

NITROUS OXIDE (N$_2$O)

Biogenic production is probably the overwhelming source of nitrous oxide (N$_2$O). Thermodynamic equilibrium calculations indicate that the equilibrium mixing ratio of N$_2$O should be about 10^{-20} (Lovelock and Margulis, 1974), some 13 orders of magnitude less than the measured N$_2$O mixing ratio of about 330 ppbv. N$_2$O may be biogenically produced by soil bacteria via nitrification and/or denitrification, depending on the soil conditions (Goreau et al., 1980; Lipschultz et al., 1981; Anderson and Levine, 1986, 1987). Nitrous oxide is chemically inert in the troposphere and has a characteristic atmospheric residence time of about 150 years. Nitrous oxide is destroyed in the stratosphere by photolysis and by reactions with excited atomic oxygen [O(^{1}D)], according to the following reactions:

$$N_2O + h\nu \rightarrow N_2 + O(^1D); \lambda \leq 337 \text{ nm} \tag{44}$$

$$N_2O + O(^1D) \rightarrow N_2 + O_2 \tag{45}$$

$$N_2O + O(^1D) \rightarrow 2 \text{ NO} \tag{46}$$

Reaction (44) is the major loss mechanism for N$_2$O and, with reaction (46), the major source of NO in the stratosphere. The catalytic nitrogen oxide cycle that leads to the destruction of O$_3$ in the stratosphere can be expressed as:

$$NO + O_3 \rightarrow NO_2 + O_2 \tag{40}$$

$$NO_2 + O \rightarrow NO + O_2 \tag{47}$$

NO is also formed by the photolysis of NO$_2$:

$$NO_2 + h\nu \rightarrow NO + O \tag{20}$$

Reactions (40) and (47) result in a net cycle of:

$$O_3 + O \rightarrow 2 O_2 \text{ (net cycle)} \tag{48}$$

Nitrous oxide absorbs in the thermal infrared at 7.5 μm in the "atmospheric window" and, hence, affects the climate of the Earth (Wang et al., 1976).

Since all the destruction of N$_2$O occurs in the stratosphere via reactions (44) to (46), N$_2$O exhibits a constant mixing ratio with altitude within the global troposphere. Recent measurements indicate that global concentrations of N$_2$O may be increasing with time, at a rate of about 0.2 percent per year (Weiss, 1981). While biogenic production appears to be the major source of N$_2$O, the combustion of fossil fuel and/or agricultural activity may be responsible for the apparent secular increase of this species (Weiss, 1981).

NITROGEN (N_2) AND CARBON DIOXIDE (CO_2)

While the bulk of atmospheric nitrogen (78% by volume) and carbon dioxide (about 340 ppmv) probably resulted from volcanic activity in the early history of our planet (the average composition of volcanic gases is H_2O (79.31%), CO_2 (11.61%), SO_2 (6.48%), and N_2 (1.29%) (Walker, 1977)), both these gases are also formed biogenically. Nitrogen is formed via denitrification and carbon dioxide is formed as a respiration and metabolic product. Both N_2 and CO_2 are chemically inert in the atmosphere. However, very small amounts of N_2 are "fixed" into NO by the action of atmospheric lightning (for example, see Levine et al., 1981, 1984). As already discussed, NO is photochemically transformed to HNO_3 and HNO_2, and is eventually returned to the biosphere in the form of nitrates and nitrites, respectively. There does not appear to be any significant atmospheric sink for CO_2. The oceans are probably the major sink for this gas (Woodwell et al., 1978; Broecker et al., 1979). Atmospheric CO_2 has increased from about 280 ppmv to 300 ppmv around 1880 to about 335 ppmv to 340 ppmv in 1980 (Siegenthaler and Oeschger, 1978), mainly due to the burning of fossil fuels (at a given location CO_2 exhibits a strong seasonal variation of several ppmv due to its uptake via photosynthesis in spring and summer). Carbon dioxide absorbs thermal infrared radiation in the atmospheric window (7 μm to 14 μm). Various theoretical radiative-temperature calculations indicate a global temperature increase of between 2° C to 3.5° C for a doubling of the present level of atmospheric CO_2, with a strong amplification (8° C to 10° C warming) in the polar areas (Siegenthaler and Oeschger, 1978).

SUMMARY

Since the early history of our planet and continuing to the present time there has been and still is a strong coupling between the atmosphere and the biosphere. Over geological time, biogenic gases have impacted the composition and chemistry of the atmosphere, as well as the climate of our planet. The biogenic production of oxygen by photosynthetic organisms transformed the oxidation state of the early atmosphere. Biogenic oxygen led to the photochemical production of ozone, which shields the biosphere from biologically lethal solar ultraviolet radiation. The photolysis of ozone leads to the photochemical production of the hydroxyl radical, which chemically transforms almost every biogenic species in the atmosphere.

Biogenic activity is the overwhelming source of atmospheric oxygen, nitrous oxide, ammonia, methane, and the reduced sulfur species, as well as an important source of carbon monoxide, nitric oxide, the reduced halogen species, carbon dioxide, and nitrogen. Once in the atmosphere, carbon dioxide, nitric oxide, and the reduced sulfur species are chemically transformed into carbonic, nitric, nitrous, and sulfuric acids. Biogenic ammonia forms the only gaseous basic constituent of the atmosphere and neutralizes the acidic environment of the atmosphere, clouds, and precipitation. Methane, the reduced sulfur species, carbon monoxide, and nitric oxide impact the chemistry of the troposphere. Nitrous oxide controls the destruction of ozone in the stratosphere. Carbon dioxide, methane, nitrous oxide, and ammonia affect the climate of our planet via their absorption of surface-emitted thermal infrared radiation. Once in the atmosphere, ammonia, nitric oxide, and the reduced sulfur species are chemically transformed and returned to the biosphere in the form of particulate ammonium, nitrate, nitrite, and sulfates, thereby completing the biogeochemical cycle. If it were not for the strong coupling between the atmosphere and the biosphere, the composition and chemical evolution of the atmosphere would have been very different.

REFERENCES

Anderson, I.C. and J.S. Levine. 1986. Relative rates of nitric oxide and nitrous oxide production by nitrifiers, denitrifiers, and nitrate respirers. *Applied and Environmental Microbiology* **51**:938–945.

Anderson, I.C. and J.S. Levine. 1987. Simultaneous field measurements of biogenic emissions of nitric oxide and nitrous oxide. *Journal of Geophysical Research* **92**:9654–9976.

Augustsson, T.R. and J.S. Levine. 1982. The effect of isotropic multiple scattering and surface albedo on the photochemistry of the troposphere. *Atmospheric Environment* **16**:1373–1380.

Berkner, L.V. and L.C. Marshall. 1965. On the origin and rise of oxygen concentration in the Earth's atmosphere. *Journal of Atmospheric Science* **22**:225–261.

Broecker, W.S., T. Takahashi, H.J. Simpson, and T.-H. Peng. 1979. Fate of fossil fuel carbon dioxide and the global carbon budget. *Science* **206**:409–418.

Canuto, V.M., J.S. Levine, T.R. Augustsson, and C.L. Imhoff. 1982. Ultraviolet radiation from the young Sun and levels of oxygen and ozone in the prebiological paleoatmosphere. *Nature* **296**:816–820.

Chameides, R.L., S.C. Liu, and R.J. Cicernone. 1977. Possible variations in atmospheric methane. *Journal of Geophysical Research* **82**:1795–1798.

Craig, H. and C.C. Chou. 1982. Methane: The record in polar ice cores. *Geophysical Research Letters* **9**:1221–1224.

Crutzen, P.J. 1979. The role of NO and NO_2 in the chemistry of the troposphere and stratosphere. *Annual Review of Earth and Planetary Science* **7**:443–472.

Ehhalt, D.H. and V. Schmidt. 1978. Sources and sinks of atmospheric methane. *Pure and Applied Geophysics* **116**:452–464.

Galbally, I.E. and C.R. Roy. 1978. Loss of fixed nitrogen from soils by nitric oxide exhalation. *Nature* **275**:734–735.

Goreau, T.J., W.A. Kaplan, S.C. Wofsy, M.B. McElroy, F.W. Valois, and S.W. Watson. 1980. Production of NO_2^- and N_2O by nitrifying bacteria at reduced concentrations of oxygen. *Applied and Environmental Microbiology* **40**:526–532.

Graedel, T.E. 1979. The kinetic photochemistry of the marine atmosphere. *Journal of Geophysical Research* **84**:273–286.

Graedel, T.E. 1980. Atmospheric photochemistry. In: *The Handbook of Environmental Chemistry*, Vol. 2, part A, O. Hutzinger, ed. Springer-Verlag, New York, pp. 107–143.

Graedel, T.E. 1985. The photochemistry of the troposphere. In: *The Photochemistry of Atmospheres: Earth, The Other Planets, and Comets*, J.S. Levine, ed., Academic Press, Inc., pp. 39–76.

Graedel, T.E. and J.E. McRae. 1980. On the possible increase of the atmospheric methane and carbon monoxide concentrations during the last decade. *Geophysical Research Letters* **7**:977–979.

Harriss, R.C. and D.I. Sebacher. 1981. Methane flux in forested freshwater swamps of the Southeastern United States. *Geophysical Research Letters* **9**:1002–1004.

Hoell, J.M., Jr., J.S. Levine, T.R. Augustsson, and C.N. Harward. 1982. Atmospheric ammonia: Measurements and modeling. *AIAA Journal* **20**:88–95.

Khalil, M.A.K. and R.A. Rasmussen. 1984. Carbon monoxide in the Earth's atmosphere: Increasing trend. *Science* **224**:54–56.

Kuhn, W.R. 1985. Photochemistry, composition, and climate. In: *The Photochemistry of Atmospheres: Earth, The Other Planets, and Comets*, J.S. Levine, ed., Academic Press, Inc., pp. 129–163.

Levine, J.S. 1982. The photochemistry of the paleoatmosphere. *Journal of Molecular Evolution* **18**:161–172.

Levine, J.S. 1985. The photochemistry of the early atmosphere. In: *The Photochemistry of Atmospheres: Earth, The Other Planets, and Comets*, J.S. Levine, Ed., Academic Press, Inc., pp. 3–38.

Levine, J.S. and F. Allario. 1982. The global troposphere: Biogeochemical cycles, chemistry, and remote sensing. *Environmental Monitor. Assess* **1**:263–306.

Levine, J.S., T.R. Augustsson, I.C. Anderson, J.M. Hoell, and D.A. Brewer. 1984. Tropospheric sources of NO_x: Lightning and biology. *Atmospheric Environment* **18**:1797–1804.

Levine, J.S., T.R. Augustsson, and J.M. Hoell. 1980. The vertical distribution of tropospheric ammonia. *Geophysical Research Letters* **7**:316–320.

Levine, J.S., T.R. Augustsson, and M. Natarajan. 1982. The prebiological paleoatmosphere: Stability and composition. *Origins of Life* **12**:245–249.

Levine, J.S., C.P. Rinsland, and G.M. Tennille. 1985. The photochemistry of methane and carbon monoxide in the troposphere in 1950 and 1985. *Nature* **318**:254–257.

Levine, J.S., R.S. Rogowski, G.L. Gregory, W.E. Howell, and J. Fishman. 1981. Simultaneous measurements of NO_x, NO, and O_3 production in a laboratory discharge. Atmospheric implications. *Geophysical Research Letters* **8**:357–360.

Lipschultz, F., O.C. Zafiriou, S.C. Wofsy, M.B. McElroy, F.W. Valois, and S.W. Watson. 1981. Production of NO and N_2O by soil nitrifying bacteria. *Nature* **294**:641–643.

Logan, J.A., M.J. Prather, S.C. Wofsy, and M.B. McElroy. 1981. Tropospheric chemistry: A global perspective. *Journal of Geophysical Research* **86**:7210–7254.

Lovelock, J.E. 1979. *Gaia: A New Look at Life on Earth.* Oxford University Press, Oxford, 157 pp.

Lovelock, J.E. and L. Margulis. 1974. Atmospheric homeostasis by and for the biosphere: The Gaia Hypothesis. *Tellus* **26**:1–10.

Margulis, L. and J.E. Lovelock. 1974. Biological modulation of the Earth's atmosphere. *Icarus* **21**:471–489.

Margulis, L. and J.E. Lovelock. 1978. The biota as ancient and modern modulator of the Earth's atmosphere. *Pure and Applied Geophysics* **116**:239–243.

Rasmussen, R.A. and M.A.K. Khalil. 1981. Atmospheric methane (CH_4): Trends and seasonal cycles. *Journal of Geophysical Research* **86**:9826–9832.

Rinsland, C.P. and J.S. Levine. 1985. Free tropospheric carbon monoxide concentrations in 1950 and 1951 deduced from infrared total column amount measurements. *Nature* **318**:250–254.

Rinsland, C.P., J.S. Levine, and T. Miles. 1985. Concentration of methane in the troposphere deduced from 1951 infrared solar spectra. *Nature* **318**:245–249.

Siegenthaler, U. and H. Oeschger. 1978. Predicting future atmospheric carbon dioxide levels. *Science* **199**:388–395.

Turco, R.P. 1985. The photochemistry of the stratosphere. In: *The Photochemistry of Atmospheres: Earth, The Other Planets, and Comets,* J.S. Levine, ed. Academic Press, Inc., pp. 77–128.

Walker, J.C.G. 1977. *Evolution of the Atmosphere.* Macmillan, New York, 318 pp.

Walker, J.C.G. 1978. Oxygen and hydrogen in the primitive atmosphere. *Pure and Applied Geophysics* **116**:222–231.

Wang, W.C., Y.L. Young, A.A. Lacis, T. Mo, and J.E. Hansen. 1976. Greenhouse effects due to man-made perturbations of trace gases. *Science* **194**:685–690.

Weiss, R.F. 1981. The temporal and spatial distribution of tropospheric nitrous oxide. *Journal of Geophysical Research* **86**:7185–7195.

Woodwell, G.M., R.H. Whittaker, W.A. Reiners, G.E. Likens, C.C. Delwiche, and D.B. Botkin. 1978. The biota and the world carbon budget. *Science* **199**:141–146.

Zimmerman, P.R., J.P. Greenberg, S.O. Wandiga, and P.J. Crutzen. 1982. Termites: A potentially large source of atmospheric methane, carbon dioxide, and molecular hydrogen. *Science* **218**:563–565.

IV REMOTE SENSING OF VEGETATION

JOHN E. ESTES AND MICHAEL J. COSENTINO

INTRODUCTION

Remotely sensed imagery provides a tool for inventory and mapping of terrestrial vegetation and land cover patterns. It literally provides a new view of the Earth's surface. Patterns at this macroscale are more easily explored and modeled when synoptic, wall-to-wall remotely sensed data is used in conjunction with detailed, finer resolution ground data and spatially located point measurement. The ability of remote sensing to map and inventory terrestrial vegetation on a global scale is a key to study of the biosphere. Besides itself, natural vegetation represents other physical conditions present within a given area (Kuchler, 1973). Soils, terrain, climate, current and historical land use, and land management practices are all manifest in "terrestrial vegetation." In order to better understand the role of natural vegetation in global ecology, it is necessary to first put the object of investigation, the globe, into context.

The need for accurate baseline data on the type and status of land cover for large areas of the globe for ecological modeling has been recognized by a number of leading scientists [e.g., Global Habitability Science Issues Document, (NASA 1983)]. Global land use/land cover data will be important in attaining insights into a variety of climatological, hydrologic, biogeochemical, and biological processes that have profound ecological implications. Terrestrial biota greatly affect the energy budget, hydrologic cycle, and biogeochemistry of the Earth, and are in turn affected by these processes. Thus, information on the terrestrial biota in the form of estimates of the spatial distribution, areal extent, and temporal dynamics of vegetation and other land cover for the globe and measures of biophysical conditions of the land, such as biological productivity (rate of accumulation of biomass), is crucial to the study of global biophysical processes. Also, understanding the effects of human impact on the biosphere requires a greatly improved understanding of the spatial and temporal dynamics of terrestrial vegetation and of human-induced changes in land cover (e.g., deforestation, desertification, and land conversions from natural to urban or agricultural).

The areas and extents of the world's biomes are not presently known with any certainty in spite of the importance of this information for modeling natural ecosystems and human impact upon them. An estimation of the worldwide land area covered by each vegetation type is important to our understanding of the relative size and role of the biosphere in the surface chemistry of the Earth. When the area of each category is multiplied by vegetation biomass and soil carbon, obtained from measurements of carefully selected sample plots representative of a given category, then the size of the global biotic pool of living and dead material in each category can be estimated. By monitoring changes in terrestrial vegetation on a global scale through time, both ecologists and resource managers can perhaps gain richer insights into global cause and effect mechanisms.

The first photograph of nature was taken in 1827 by Joseph Nicephore Niepce. The picture consisted of a view from the window of his house in Gras, France. The photograph's exposure time was eight hours. The photographic emulsion was bitumen of Judea coated on a metal plate (Estes, 1969). The speed of photographic emulsions used in aerial imaging today have increased by well over four million times (Lo, 1976). Similar advances have been made in sensor systems from simple cameras to complex multispectral array detectors and in sensor platforms from tethered balloons to satellites. Applications of remote sensing have advanced from mapping German forests using a camera in a hot air balloon in 1887 (Spurr, 1960) to the use of the Advanced Very High Resolution Radiometer (AVHRR) on the National Oceanic and Atmospheric (NOAA)–6 satellite for the production of a global vegetation index by Compton J. Tucker at the National Aeronautics and Space Administration's Goddard Space Flight Center (GSFC).

The following material presents a brief historical perspective on these developments. The chapter is divided into two sections. The first section discusses the development of remote sensing systems and platforms, and contains some discussion of measurement and analytical equipment and interpretation procedures. The division between aircraft and space platforms is depicted in Figures IV–1 and IV–2. The second major section of this chapter deals with the development of the applications of aerial imagery. The focus of the latter section is the applications of remote sensing to the mapping, inventory, monitoring, and modeling of surface cover information, particularly vegetation. Figure IV–3 represents a synopsis of the material presented in this section. The material presented here is highly selective. Should the reader desire more background, particularly on the systems and platforms section, he or she should consult the following references: Quakenbush (1960); McCamy (1960); Land (1959); Koher and Howell (1963); Clark (1967); Estes (1969); Reeves (ed.) (1975); and Powers and Gengry (1979), and the *Manual of Remote Sensing*, second edition, Volume I (Colwell, 1983). For further information on applications, the reader is directed to the American Society of Photogrammetry's *Manual of Remote Sensing*, second edition, Volume II (Colwell, 1983). We conclude with some thoughts on the future applications of remote sensing to the study of the biosphere and a summary of remote sensing of global vegetation.

SYSTEMS AND PLATFORMS

Even in Aristotle's time, sunlight was known to be indispensable to life. Aristotle understood the principle of the camera obscura and philosophized at length on the nature of light. In a camera obscura, light passing through a small hole or aperture in a dark box or chamber can be made to form a picture. Ibn al-Haltam (Alhazen of Basra), an Arab scholar of the eleventh century, and the Hebrew scholar Levi ben Gerson both recorded accounts of the use of pinhole cameras to observe eclipses. The pioneering Leonardo da Vinci was the first to describe light passing through a small opening into a darkened room where it formed an image on a flat surface.

Joseph Nicephore Niepce and Louis Jacques Mande Daguerre are generally agreed upon as being the coinventors of the first practical means of photography. Niepce is credited with taking what is considered to be the world's first true picture without an engraving stage. This was in 1827. After agreeing to collaborate, Niepce and Daguerre found their greatest problem to be fixing an exposed image to make it resistant to the further action of light. Herschel's 1819 discovery of sodium thiosulphate ("hypo") was not known by Daguerre until 1839. On August 19, 1839, six years after Niepce's death, Daguerre announced the invention of the Daguer-

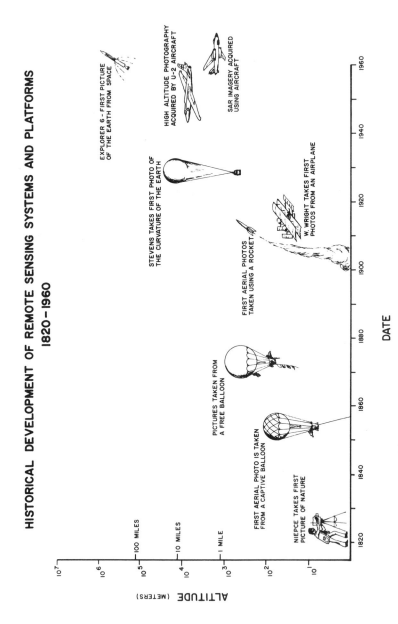

FIGURE IV-1 Historical development of remote sensing systems and platforms 1820–1960.

78

HISTORICAL DEVELOPMENT OF REMOTE SENSING SYSTEMS AND PLATFORMS
1959 - 1990s

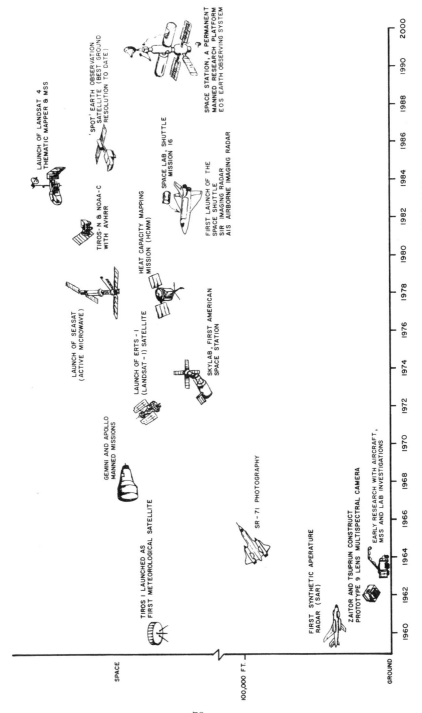

FIGURE IV-2 Historical development of remote sensing systems 1959–1990s.

79

HISTORICAL DEVELOPMENT OF REMOTE SENSING FOR ECOLOGICAL ANALYSIS

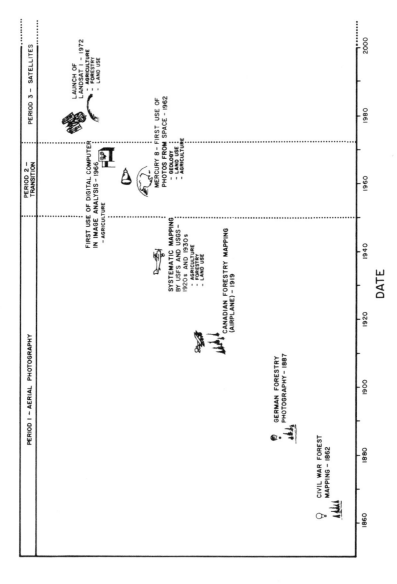

FIGURE IV–3   Historical development of remote sensing for ecological analysis.

80

reotype in a report on the work he and Niepce had done. The French government promptly bought the invention.

In 1841, William Henry Fox Talbot accidentally discovered that an image formed on paper coated with silver iodide, though barely visible, could be strengthened or developed with gallic acid and silver nitrate. Talbot's invention of the Calotype and later Talbotype process made it possible to produce any number of positive copies from the negatives produced.

The first compact portable camera with folding bellows was described by Pierre Sequier in 1840. This was the same year that the first suggestion of acquiring aerial photography (from balloons) appeared in a cartoon in France entitled "Daguerrotypomania." In 1858, Gaspard Felix Tournachon, planning to produce a topographic map from photographs, ascended in a captive balloon to an altitude of several hundred meters and took pictures of Paris. Colonel Aime Laussedat, often credited as the father of modern photogrammetry, followed this in 1858 with experiments involving a glass-plate camera supported by several kites. Laussedat finally succeeded in developing the mathematical formulas for converting overlapping perspective views into an orthophotographic projection on any plane. This work was not reported until 1898. During the period between 1860 and 1900, advances were made in lens systems for cameras. These extended the sensitivity of photographic emulsions to larger wavelengths that could be seen with the unaided eye and included techniques for producing sharper images. In 1879, photographs were taken from an untethered balloon over Paris by Triboulet. Yet balloons and kites are not navigable platforms and, although many useful pictures were acquired without piloted or controlled platforms, there was little advance in the use of aerial photography.

In 1906, an engineer named Albert Maul demonstrated that a rocket could be used to obtain photographs. The rocket, propelled by compressed air, rose to an altitude of 2,625 feet. This was three years before Wilbur Wright took the first recorded photographs from an airplane. On April 24, 1909, Wright took motion pictures over Centocelli, Italy. The first aircraft photography used for mapping was described by Tardivo in 1913 in a paper presented at the meeting of the International Society of Photogrammetry in Vienna.

World War I produced the first practical use of aerial photography. Its potential for military applications was recognized almost immediately. By the end of 1915, Lt. Col. J.T.C. Moore Brabazon in collaboration with Thornton Pickard Ltd. had designed and produced the first aerial camera. Improved models of this camera convinced the military of the value of aerial reconnaissance and photointerpretation. Following World War I there was a growth of commercial aerial survey companies in the 1920s and 1930s. In 1934, the first issue of the scientific journal *Photogrammetric Engineering and Remote Sensing* was published. Other developments between World Wars I and II included the patenting of multilayered film in 1924 by Mannes and Godousky. This patent eventually led to the marketing of Kodachrome film in 1935. Also during this time, Captain Albert W. Stevens took the first photographs showing the curvature of the Earth. This was accomplished in 1936 as part of a project sponsored by the U.S. Army and National Geographic. The photo was taken from an untethered balloon at an altitude of 72,395 feet over South Dakota. It was this series of stratospheric balloon flights that prompted Stevens to develop a black-and-white infrared-sensitive film in 1941. The film was needed to provide sharper images of surface features blurred due to Rayleigh scattering. Since energy in the shorter wavelength, blue portion of the spectrum is scattered more than energy in the longer wavelength, red and infrared portions of the spectrum, films with sensitivity to longer wavelengths can provide clearer images of surface objects from high altitudes. This is why standard commercial panchromatic aerial photography for many years has been acquired with a filter that removes the blue light.

The first suggestion to provide color processing for infrared film was made in 1941. Researchers at Kodak Laboratories in England reported on the possible use of color infrared film for detecting military camouflage. After enlarging on the early experiments in 1942, Spencer, Marriage, Jelly, and Wilder patented the basic structure used in the process. The patent was issued in 1946.

World War II gave great impetus to developments in photo-interpretation. In 1946, after the war, the first photographs from space were obtained from captured German V2 rockets. Thermal infrared imaging systems were tested out of Wright Patterson Air Force Base using systems essentially developed by the Germans during the war but not put into widespread use. The first side-looking airborne radar was developed in a crash program beginning in 1954 by Westinghouse Corporation. The first photos of the Earth from an orbiting satellite were acquired by Explorer 6 in 1959.

To provide cold war intelligence, the U-2 was designed as a photographic platform that could fly well above the range of known enemy rockets. In August 1955, the first model of the U-2 was flown. By the time of the first orbiting satellites in the late 1950s, a wide array of sensor systems and platforms had been employed for collecting aerial photography and data from other sensor systems. In addition, the developments in the area of computers began to make digital processing of remotely sensed data possible for the first time. This new field of scientific and technological research and development produced a proliferation of systems that employed nonphotographic techniques to create digital images of reflected, emitted, and transmitted energy from surface and near-surface objects and phenomena. This led to the coining of the term *remote sensing* in the early 1960s.

After Explorer, a series of satellites from the TIROS series meteorological satellites through the Mercury, Gemini, and Apollo series satellites helped pave the way for the launch in 1972 of the first Earth Resources Technology Satellite ERTS-1, later renamed Landsat-1. This launch followed an intense period of research in the 1960s on the use of multispectral scanners for acquiring remotely sensed data. These scanners recorded electromagnetic energy in selected wavelength bands in digital form. These scanning systems employ mirrors and detectors to acquire data one line at a time to produce an image.

Since Landsat-1, we have had Earth resources sensors systems, which were designed to collect thermal infrared and synthetic aperture radar data from space. Although early meteorological satellites carried thermal infrared sensor systems, the Heat Capacity Mapping Mission (HCMM) for thermal infrared and Seasat for radar were both launched in 1978.

The advanced multispectral scanner, Thematic Mapper, was launched on Landsat-4 in 1980. Table IV–1 presents a comparison of the multispectral scanner system on Landsats-1,-2 and -3, and the Thematic Mapper on Landsat-4.

In 1981, the first of the meteorological satellites carrying the Advanced Very High Resolution Radiometer (AVHRR) was launched. Table IV–2 details the characteristics of this sensor system.

Planning and development of future satellite and sensor systems is well under way. France launched SPOT (System Probaterie de la Observation Terre), a broad-band, 10-meter to 20-meter spatial resolution linear array, multispectral system in 1986. Table IV–3 presents the characteristics of the SPOT High Resolution Visible (HRV) Instruments.

The Japanese and European space agencies also have advanced multispectral scanner and synthetic aperture radar systems planned for launch in the 1980s. The United States has and will continue to use the Shuttle to test new sensor systems (e.g., the shuttle imaging radar experiments). These experiments involve the use of multifrequency radars with varying look angles. In addition, plans are under way to fly high spectral resolution systems on aircraft and spacecraft, such as the Advanced Imaging Spectrometer, which images 128 spectral channels in the near-infrared portions of the electromagnetic spectrum.

Finally, planning is well under way on a package of sensor systems to fly

TABLE IV-1　SPECTRAL PROPERTIES OF THE LANDSAT SENSORS (THEMATIC MAPPER, MULTISPEED SCANNER)

	Thematic Mapper (TM)		Multispectral Scanner Subsystem (MSS)	
	Micrometers	Sensitivity (NEΔρ)	Micrometers	Sensitivity (NEΔρ)
Spectral Band 1	1.45–0.52	0.8%	0.5–0.6	0.57%
Spectral Band 2	0.52–0.60	0.5%	0.6–0.7	0.57%
Spectral Band 3	0.63–0.69	0.5%	0.7–0.8	0.65%
Spectral Band 4	0.76–0.90	0.5%	0.8–1.1	0.70%
Spectral Band 5	1.55–1.75	1.0%		
Spectral Band 6	10.40–12.50	0.5 K (NEΔT)		
Spectral Band 7	2.08–2.35	2.4%		
Ground 1FOV		30 m (Bands 1–6) 120 m (Band 7)	82 m (Bands 1–4)	
Due rate		85 Mbps	15 Mbps	
Quantization levels		256	64	
Weight	258 kg	68 kg		
Size		1.1 × 0.7 × 2.0 m	0.35 × 0.4 × 0.9 m	
Power	332 watts	50 watts		

TABLE IV-2　ADVANCED VERY HIGH RESOLUTION RADIOMETER (AVHRR)
AVHRR/2—Five Channel Instrument (3)

1.　0.58 –1.68 μm
2.　0.725– 1.10 μm
3.　3.55 – 3.93 μm
4.　10.3 –11.3　μm
5.　11.5 –12.5　μm

	Channels				
Characteristics	1	2	3	4	5
Spectral Range (μm)	0.58–0.68	0.725–1.1	3.55–3.93	10.3–11.3	11.5–12.5
Detector	Silicon	Silicon	InSb	HgCdTe	HgCdTe
Resolution (km)	1.1	1.1	1.1	1.1	1.1
1FOV (mrad)	1.3	1.3	1.3	1.3	1.3
NETD @ 300 K	—	—	0.12	0.12	0.12
S N 0.5% albedo	>3:1	>3:1	—	—	—
MTF (1FOV/single bar)	0.3	0.3	0.3	0.3	0.3

Optics:　8-inch diameter a focal cassegrainian telescope
Scanner:　360-rpm hysteresis synchronous motor
Cooler:　2-stage passive

In-orbit data obtained after completion of the protoflight instrument has shown the necessity of eliminating spectral overlap with channel 2 if snow-cover areal extent is to be accurately measured.

TABLE IV-3 CHARACTERISTICS OF SPOT HIGH RESOLUTION
VISIBLE (HRV) INSTRUMENTS

Characteristics of the HRV Instruments	Multispectral Mode	Panchromatic Mode
Spectral Bands	0.50–0.50 μm 0.61–0.68 0.79–0.89	0.5–0.73 μm
Instrument Field of View	4.13 deg	4.13 deg
Ground Sampling Interval (Nadir Viewing)	20 m × 20 m	10 m × 10 m
Number of Pixels per Line	3,000	6,000
Ground Swath Width (Nadir Viewing)	60 km	60 km

in space in the era of space stations in the 1990s. NASA planning for this effort was conducted under the name of System Z, later renamed EOS, for Earth Observing System. EOS is envisioned by NASA to fly in polar orbit as part of the space station complex. EOS is being designed to address potential science issues of the 1990s through the information synergism created by the analysis of data from a number of sensor systems operating simultaneously from near-ultraviolet to microwave wavelengths (Table IV–4). Planning for EOS also includes the implementation of an advanced user interactive distributed surface data system to insure efficient use of the data generated.

Systems and platforms for the acquisition of remotely sensed data have seen considerable advances in technological capability in the last 125 years. And there has been an increase in potential applications of these systems and platforms. Developments within the area of ecological analysis are described in the following section.

APPLICATIONS OF REMOTE SENSING

We divide applications of remote sensing to ecological analysis into three periods. The first, the period before 1950, constitutes the time of the initial applications of aerial photography to ecology. The second period, 1950 to 1972, represents the transition of applications from photography to unconventional imagery systems (e.g., thermal infrared scanners and side-looking airborne radars) and from low altitude aircraft to satellite platforms. From 1972 to the present, emphasis has been placed on the applications of multispectral scanner and radiometer data obtained from operational satellite platforms.

TABLE IV-4 EARTH OBSERVING SYSTMES (EOS) INSTRUMENTS

Instrument	Measurement	Spatial Resolution	Coverage
1. Automated Data Collection & Location System (ADCLS)	Data and command relay and location of remotely sited measurement devices	Location to 1 km for buoys, to 1 m for ice sheet packages	global, twice daily
	SISP—Surface Imaging & Sounding Packages		
2. Moderate Resolution Imaging Spectrometer (MODIS)	Surface and Cloud imaging in the visible and infrared .4 nm–2.2 nm, 3–5 µm, 8–14 µm resolution varying from 10 nm to .5 µm.	1 km × 1 km pixels (4 km × 4 km open ocean)	global, every 2 days during daytime plus IR nighttime
3. High Resolution Imaging Spectrometer (HIRIS)	Surface Imaging .4–2.2 nm, 10–20 nm spectral resolution	30 m × 30 m pixels	pointable to specific targets, 50 km swath width
4. High Resolution Multifrequency Microwave Radiometer (HMMR)	1–94 GHz passive microwave images in several bands	1 km at 36.5 GHz	global, every 2 days
5. Lidar Atmospheric Sounder and Altimeter (LASA)	Visible and near infrared laser backscattering to measure atmospheric water vapor, surface topography, atmospheric scattering properties	vertical resolution of 1 km, surface topography to 3 m vertical resolution every 3 km over land	global, daily atmospheric sounding; continental topography total in 5 years
	SAM—Sensing with Active Microwaves		
6. Synthetic Aperture Radar (SAR)	L, C, and X-Band Radar images of land, ocean, and ice surfaces at multiple incidence angles	30 m × 30 m pixels	200 km swath width daily coverage in regions of shifting sea ice

Instrument	Measurement	Coverage
7. Radar Altimeter	Surface topography of oceans and ice, significant wave height	10 cm in elevation over oceans	global with precisely repeating ground tracks every 10 days
8. Scatterometer	Sea surface wind stress to 1 m/s, 10° in direction Ku band radar	one sample at least every 50 km	global, every 2 days

APACM—Atmospheric Physical & Chemical Monitor

9. Doppler Lidar	Tropospheric winds to 1 m/s doppler shift in laser backscatter	1 km vertical, 2° longitude, 2° latitude	global, twice daily surface to 100 mb
10. Upper Atmosphere Wind Interferometers	Upper atmospheric winds to m/s, doppler shift in O_2 thermal emissions	3 km vertical, 2° longitude, 2° latitude	global, daily
11. Tropospheric Composition Monitors	Trace chemical constituents of the troposphere	varies from total column density to 1 km vertical, from 1° to .1° horizontal	global, daily, surface to 100 mb
12. Upper Atmosphere Composition Monitors	Trace chemical composition passive emission detectors at wavelengths from UV to microwave	3 km vertical 2° longitude, 2° latitude	tropopause to 120 km global daily day and night coverage
13. Energy and Particle Monitors	Solar Emissions from 150–400 nm, 1 nm spectral resolution. Earth radiation budget Total Solar irradiance Particles & fields environment	total solar output	roughly continuous sampling, at least twice daily for solar observations

87

BEFORE 1950

Although the development of aerial photo-interpretation followed closely the development of the camera, it was also closely associated with the development of terrestrial photogrammetry. While photo-interpretation is defined as the art and science of obtaining reliable measurements by means of photography, photo/image interpretation is to photogrammetry as statistics is to mathematics. Both image interpretation and statistics represent techniques by which probabilistic statements are made concerning objects, phenomena, and relationships in our environment. Photogrammetry and mathematics, on the other hand, deal with more precise methods of describing objects and demonstrating relationships (Estes and Simonett, 1975).

The theory of perspective drawings to determine dimensions was published by Lambert in 1759. Although the French hydrographer Beautemps-Beaupre used this theory to provide maps of the Tasmanian coast between 1791 and 1793, the merging of these perspectives to photography to produce the science of photogrammetry did not occur until the 1860s (Howard, 1970). In 1861, Col. Aime Laussedat, credited as the father of photogrammetry, applied Beautemps-Beaupre's principles to terrestrial perspective photography, which led to the first orthographic maps from photographs of a village near Versailles.

The first application of photogrammetry to forestry is recorded by Howard as 1862 and by Spurr as 1887. Howard (1970) reports that the first known ecological use of aerial photography occurred in the American Civil War. Around Richmond, Virginia, in 1862, the Union Army used photographs to delineate pinewoods, swamps, and rivers. This report by Howard is disputed, however, by Fischer (1975), who quotes an "exhaustive" study by Hayden that says " . . . no aerial photographs have been found in War Department Archives or in those of the Signal Corps."

What is agreed upon by all, however, is the use of air photos by German foresters in 1887. In that year, German foresters experimented with constructing forest maps from photographs taken from a hot air balloon. These foresters were able to distinguish a 100-year-old stand of spruce, oaks, and maples, as well as mixed plantings along brooks. The only important failing in this effort was that the scale varied and was indeterminate (Spurr, 1960). The first aerial photographs were taken by Wilbur Wright over Centocelli, Italy, in 1909, three years after Albert Maul took the first photographs from a rocket.

Modern aerial photography for peaceful purposes is said to have begun in 1913, when an Italian aircraft took photographs of the town of Benghazi and nearby landscape in Libya (Howard, 1976). These photographs were used to construct a mosaic for geological mapping. In 1914, World War I rapidly expanded the applications of aerial photography and brought on military photo-interpretation. Shortly after the beginning of hostilities, a German airship on wartime patrol was captured and found to contain an aerial camera.

Early use of air photos involved pinpointing, or the precise photographing of selected points. This changed, however, at the time of the Palestine campaigns, when strip photography began. By 1918, photographic units were providing the French Army with up to 50,000 photographs a week prior to the war's final offensive. Also in use by 1918 was the popular present-day "film-filter" combination of panchromatic plate and yellow haze filter.

After the war, aerial photographs came to be widely used, particularly in Canada, the Middle East, and the United States. As early as 1920, Thomas, writing in *Nature*, drew attention to the potential of aerial photography in archaeology, botany, geography, geology, and meteorology. During the same period, Blandford and Watson were negotiating for an aerial survey of delta forest in Burma (Kemp et al., 1925).

The first major peacetime use of airplanes and photo-interpretation in forest-stand mapping was in Canada in 1919, when Ellwood Wilson (1920) used aerial photographs for forest stock-mapping. Half a million acres were photographed at a cost of 2 to 6 cents an acre (Howard, 1970). In 1923, the principles of aerial photography for timber-volume estimation were being worked out by Hugershoff at the forest research institute at Tharaudt. By 1924, aerial photographs were being used in many parts of the British Commonwealth. For example, 1,000 square miles of the Irrawaddy Delta in Burma had been photographed vertically; aerial photos of Australia were being acquired with a 4-inch by 5½-inch format. In 1928, Bourne used aerial photographs for ecological studies in central Africa; by 1929 Sealey in Canada had carried out a forest survey in which tree heights were recorded (Howard, 1970).

In the United States in the late 1920s and early 1930s, a plan was initiated to photograph farm and ranch lands systematically throughout the entire United States. The Forest Service photographed much of the United States timber reserve and the U.S. Geological Survey began to photograph large areas of the United States to make both topographic and geologic maps (Fischer, 1975). The Tennessee Valley Authority began the large-scale use of aerial photography for land use and resource management purposes (Estes et al., 1980).

In 1931, G.W. Leeper employed aerial photographs in a soil survey of

Mount Gellibrand, Australia. In that same year, Bourne wrote of the role of aerial photographs in soil surveys in the United Kingdom. In 1936, a sea lion census was conducted in California inaugurating the use of aerial imagery for the analysis of animal populations. Since that time the use of air photos and satellite imagery for rangeland inventory has continued as the basis for all range resource inventories conducted by the United States Forest Service (Colwell, 1960). To date, caribou and muskoxen in Canada, game animals in central Africa, ducks in the Great Central Valley of California, and snow geese in Delaware have been censused using aerial photography. Thermal infrared imagery began to be employed in the 1960s for censusing range cattle in California and elephants in Africa.

1950–1972

For agriculture and forestry, the 1950s and 1960s were a period of transition from the more conventional use of black-and-white aerial photography to the use of color, color infrared, and multispectral photography and imaging systems. A classic early work in this area is the research on the use of special-purpose aerial photography for the recognition and classification of vegetation types, and for the detection of diseased and stressed vegetation (Colwell, 1956).

Although some photography from rockets was obtained in the late 1950s, the first manned satellite to acquire imagery was Mercury 8 in 1962. While these and the images collected by the later Gemini and Apollo programs were of little use in global ecological studies, they paved the way for the free-flying satellites of the 1970s. In 1961, the Committee on Remote Sensing for Agricultural Purposes was formally constituted by the National Research Council. The committee examined the potential of remote sensing to assist researchers beyond the areas of crop and pest detection (e.g., information on crop and forest production, management, and marketing from airborne or space vehicles).

In 1964, research funded by NASA and USDA began to examine problems of data reduction and discriminant analysis in the timely production of crop and forest information. By late 1965, NASA, in cooperation with USDA, Purdue University, and the University of California at Berkeley, established two laboratories to plan, coordinate, and conduct research to develop and apply aerospace remote sensing techniques to the study of agricultural crops, forests, and natural grasslands. A Laboratory for Applications of Agricultural Remote Sensing was established at Purdue and a Center for Forestry Applications was founded at Berkeley. Research conducted at these institutions through the remainder of the 1960s provided the foundations for an Earth-viewing, remote sensing capability that used satellites and digital computer processing technologies.

In 1966, the first use of digital computers for the analysis of multispectral measurements collected from aircraft were made for an agricultural field in Indiana by the Purdue LARS computer laboratory. Results of this effort demonstrated that multispectral image processing could be used to distinguish wheat from oats, even showing oats planted within a crop of wheat and scattered wheat along a stream from seeds transported downstream. During this period, a wide range of research on the spectral properties of Earth surface materials was conducted. This research was largely aimed at understanding the reflectance properties of surface materials. It was felt at the time, and to some extent still is, that given enough knowledge of the multispectral reflectance properties of these materials, sensor systems could be designed to unambiguously separate an object from its background. This is the foundation of what became known as the multispectral concept in remote sensing (Lent, 1969). These spectral reflectance studies provided the basis for the selection of the spectral bands chosen for Landsat-1.

A number of key large-scale projects undertaken in the 1970s led to significant advances and to the current capabilities in the remote sensing of vegetation. These projects included the Corn Blight Watch Experiment (CBWE), the Large Area Crop Inventory Experiment (LACIE), and the Ten Ecosystem Experiment.

In 1970, southern corn leaf blight caused extensive damage to the United States corn crop. The Corn Blight Watch Experiment (CBWE) was initiated in April 1971. CBWE was designed to employ information derived from multispectral remote sensing, digital pattern recognition techniques, and manual interpretation of color infrared aerial photographs. The primary objectives of this work were to 1) detect and control the development and spread of corn blight across the corn belt during the growing season, 2) assess levels of infection present, 3) estimate the land area affected, and 4) generalize information obtained from surface sampling to assess yield impacts. The experiment was also set up to assess the potential applicability of the techniques to similar future situations. Prior to use of these remote sensing techniques (and associated surface sampling techniques), information concerning the spread of corn leaf blight was often based on hearsay.

CBWE employed high-altitude aircraft taking color infrared photography, low-altitude aircraft collecting multispectral measurements, and ground observations. All data were collected according to a statistical sampling strategy. The experiment was a success. Location and spread of corn leaf blight was accurately monitored over large areas for the first time. Data was acquired for a seven-state area every two weeks. CBWE demonstrated that large areas could be assessed accurately and rapidly using: 1) sound statistical sampling strategies, 2) manual interpretation of small-scale, high-altitude color infrared photography, and/or 3) computer processing of multispectral scanner measurements. The experiment also produced a relative evaluation of the manual and machine-assisted remote sensing approaches and provided considerable encouragement for the use of machine-processing approaches for multispectral data. Finally, CBWE demonstrated that large amounts of data could be obtained, reduced, interpreted and used in short periods of time (MacDonald et al., 1972).

1972–PRESENT

Agriculture

The Landsat–1 Satellite was launched in July 1972. One of the first major programs that attempted to use Landsat data for the study of vegetation was the Crop Identification Technology Assessment for Remote Sensing (CITARS) experiment. This experiment was designed to evaluate a variety of existing quantitative techniques for the identification of specific crops by using satellite remote sensing. CITARS had five specific objectives. These were to:

1. determine the accuracy of Landsat data for identifying corn and soybeans;
2. assess the impact of different geographic locations with different physical characteristics and cultural practices on crop identification;
3. use machine data-processing and develop quantitative measures of variations in crop-identification accuracy;
4. test the concept of "signature extension" (that is, the extent to which classification algorithms developed for one location could be applied to other areas); and,
5. assess the benefits of multidate Landsat classification techniques (Myers, 1983).

CITARS demonstrated that multidate satellite remotely sensed data improved the potential for accurate classification of agricultural cover types. The program also pointed out two major problems that still confront researchers today. These research questions are known as the mixed pixel problem and the signature extension problem. Pixel refers to "picture element." A pixel represents the areal extent of the instantaneous field of view (IFOV) of a scanning remote sensor system. A pixel, then, defines the resolution limit of the system. When two or more types of vegetative cover are present within a given picture element (creating a mixed rather than a pure pixel), the probability of correct classification employing standard statistical pattern recognition procedures can be significantly decreased. This, in brief, is the mixed pixel problem.

Because Landsat spectral signature pixel level resolutions are essentially 80 m for MSS (multispectral scanner) and 30 m for TM (thematic mapper), there are problems in finding pure pixels in agriculture areas and even in some areas of natural vegetation. This can in some cases present a significant problem in identifying or classifying the types of vegetation in a given scene. When correlations between sensor-recorded digital values and surface cover types developed for one area are used to classify a different area, the probability of correct classification again often decreases. This, in brief, is the signature extension problem.

Results of CITARS affected the design of a more focused agricultural research program: the Large Area Crop Inventory Experiment (LACIE), which began in 1974. LACIE was designed to determine the potential to forecast harvests of a single and important crop, wheat, on a worldwide basis using satellite remote sensing technology. For the first time, biological production of a major crop was to be estimated on a global scale. LACIE produced particularly good estimates of wheat acreage in geographic areas having large field sizes (fields having rectilinear dimensions large relative to Landsat MSS pixel resolution, which is about 80 m). These areas included the hard, red winter wheat areas of the United States, the Soviet Union, and Argentina. In such areas, the large fields with a larger percentage of pure versus mixed pixels tended to improve classification accuracies.

In a quasi-operational test, LACIE in season forecasts predicted a 30 percent shortfall in the 1977 Soviet spring wheat crop that came within 10 percent of the official Soviet figures released months after harvest. In addition, LACIE midseason winter wheat forecasts predicted, within 7 percent, a 23 percent above normal Soviet winter wheat crop several months before harvest. This prediction was made with a coefficient of variation for the estimate of total wheat harvest of 3.8 percent (MacDonald and Hall, 1980).

In 1978, LACIE experiments were extended to include more types of crops as well as forests and rangelands. This new program was called Agriculture and Resource Inventory Surveys Through Aerospace Remote Sensing (AgRISTARS). AgRISTARS was a joint project between NASA and USDA. Crop assessment was being enhanced in several cases by the construction of agrometeorological models and canopy reflectance models.

World Meteorological Organization weather data was determined to be a particularly valuable source of information supporting use of remote sensing for modeling crop yields.

Agricultural research in the 1960s and 1970s demonstrated that timely, large-scale agricultural resource surveys are now feasible. Remotely sensed imagery must be supported by collateral information with appropriate spatial and temporal characteristics. Further data processing hardware and software support integrated and operated by skilled personnel must also be available for both digital and analog imagery.

The Corn Blight Watch Experiment, CITARS, LACIE, and AgRISTARS led to major advances in machine-assisted processing of remote sensing imagery of vegetation, the development of vegetation canopy reflectance models (models of canopy structure and of the reflection), knowledge of the absorption and emission of electromagnetic radiation by vegetation canopies, and recognition of the importance of models of the energy exchange properties of vegetation. With regard to the latter, these programs demonstrated that a knowledge of crop type, crop phenology, and starting conditions (including time of planting or emergence of vegetation, and soil moisture content at that time) can provide a strong basis for yield production.

In essence, if we know what is planted where and when, under what environmental conditions, and have some information on how the crop was produced in the past, we can predict with reasonable accuracy a projected yield. Data from Landsat and meteorological satellite sensor systems can then be used during the growing season to verify whether conditions are continuing to produce the expected yield or if there has been a deviation in conditions. Should such deviations occur, adjustments in yields can be made again based on knowledge of the physiological response of the crop to similar past conditions. The major challenge beyond this level is to predict variation away from the expected yield, especially to predict major crop surpluses or shortfalls, and to be more geographically precise in our ability to detect specific types of stresses on crops and the effects of these stresses.

Wetlands

Wetlands are extremely important ecosystems. Although they cover only a small fraction of the globe, they play a major role in global biogeochemical cycling. Landsat surveys of wetlands have been conducted by various investigators. In addition, a number of generalized land use and vegetation mapping projects have included wetland categories. Klemas et al. (1975)

studied Delaware wetlands using Landsat and Skylab imagery. Carter et al. (1977) describe a United States Geological Survey project involving Landsat imagery to study the Great Dismal Swamp of Virginia and North Carolina. It was found that no single date of Landsat imagery could provide adequate information for vegetation classification in the Great Dismal Swamp area. However, vegetation types could be identified by making seasonal and temporal comparisons using February and April imagery (Carter et al., 1977; Gammon and Carter, 1976). Yet general classification accuracies remain relatively low.

Ernst and Hoffer (1979) attempted to improve the accuracy of an inland wetland classification using ancillary data, such as soil information, in an appropriate classification algorithm. The researchers identified open water and five homogeneous wetland habitat types (deep marsh, shallow marsh, scrub swamp, hardwood swamp, and tamarack) using a discriminant function analysis and found that the overall classification performance was 68 percent. That is, based on an analysis of the data, 68 percent of the pixels in the scene processed were determined to be the vegetation type in which they were classified. Conventional machine-assisted classification results showed that certain wetland types could be identified by their spectral characteristics, but some confusion did occur within specific wetland types and between wetland and upland cover types (Ernst and Hoffer, 1979, 1981). Where maximum likelihood classification algorithm, employing a multistage decision logic called a "layered classifier" (Swain, 1979), was used to combine soil data with the Landsat MSS, data classification performance improved significantly. In addition, hardwood swamps could be identified using the combined soils/spectral data. Use of the layered classifier also improved the accuracy of identification of shrub swamps and shallow mashes but the overall performance was still somewhat low, 84.3 percent compared to 71.1 percent for the single-stage classification (Ernst and Hoffer, 1979, 1981).

Some studies have attempted to estimate wetland biomass. Carter (1976) used measurements of the areal extent of major wetland species interpreted from Landsat data plus the typical values for production for marsh area in Virginia to estimate wetland biomass. Bartlett and Klemas (1980) studied tidal wetlands in Delaware, which were dominated by *Spartina alterniflora*. For this study, radiance measurements and ecological field data, including field biomass, were collected for vegetation plots. Spectral reflectance of *Spartina alterniflora* was measured in the four Landsat MSS bands. The paired radiometric and biotic measurements for each plot were then used in regression analysis to identify significant relationships. The relationships were tested using Suit's (1972) model, which determines composite canopy reflectance from the optical properties of individual canopy components and the geometry of those components within the canopy. Results obtained by Bartlett and Klemas (1980) show that canopy reflectance measurements in bands four and five (visible reflectance) are inversely related to the percentage of green vegetation within the canopy. Green biomass relationships were somewhat weaker in the infrared but high correlation of ratioed reflectance in band seven/band five (reflectance infrared to red) with the green biomass of *S. alterniflora* was observed ($r = 0.85$). This relationship appears to be linear over a wide range of green biomass values (i.e., 2 g to 1,000 g dry wt/m^2). Bartlett and Klemas (1980) then tested this relationship at three test sites. Based on this test, these researchers concluded that spectral measurements made from sensors such as Landsat have the potential for providing information on the standing crop and emergent biomass for selected wetlands plants and that the quantitative assessment of marshes dominated by *S. alterniflora* using satellite data appears plausible (Lulla, 1983).

Forest and Rangeland

Digital and analog remote sensing from both aircraft and satellite platforms have also been applied to large-scale forest and range inventory and monitoring (Heller and Ulliman, 1983; Carneggie et al., 1983).

Wildland and rangelands have been analyzed using Landsat data both in digital and visual format (Carneggie et al., 1975). Haas et al. (1975) developed several indexes to monitor vegetation conditions and biomass in the Great Plains and southwestern rangelands. The use of Landsat for monitoring erosion in southern Arizona has been demonstrated by Morrison and Cooley (1973), who found that band five provided the best definition of modern arroyos. Bentley et al. (1976) made essentially similar observations in assessing ephemeral and perennial rangeland vegetation and soils using Landsat data. An example that extends these mapping applications is the report by Johnston (1978). In this work, Johnston found that densitometric processing of Landsat MSS bands five, six, and seven can be used to identify rangeland productivity classes. Finally, employing an information systems approach, researchers at the University of California at Santa Barbara have achieved an overall classification accuracy of 89.9 percent in a pilot study demonstrating the potential of Landsat to map chaparral brush fuel types and densities. The geographic information system employed in this study consisted of registered data layers containing fire history, rainfall, soils, terrain, vegetation, and insolation data.

While other researchers have examined the problems associated with the use of satellite remote sensing for mapping potential fuel localities, machine-assisted approaches have failed to produce sufficient classification accuracies. In a study in the Lolo National Forest in western Montana, Shasby et al. (1981) achieved classification accuracies of only 68 percent with a sequential classifier that incorporated Landsat and a fuels class terrain model.

With respect to forestry, the joint USDA/U.S. Forest Service/NASA Ten Ecosystem Study, initiated in 1976, demonstrated that Landsat MSS data with appropriate machine-assisted processing techniques can distinguish hardwood, softwood, grassland, and water and make inventories of these classes with an accuracy of 70 percent or greater at an operation cost of 11 cents per hectare (Mazade, 1981). If general classification levels, such as forest and grassland, were sufficient, 90 percent accuracies were obtained. The primary objective of the Ten Ecosystem Study, which examined 630,000 hectares in ten selected ecosystems, was to explore the feasibility of using Landsat multispectral data and automated pattern recognition analysis to inventory forest and grassland resources. The study, which divided the continental United States into ten broadly defined ecological classes and examined the similarities and differences among them, built on the results of many local research projects and has served as a prelude to larger scale investigations. It was one of the major studies conducted under the U.S. Forest Service's National Forestry Applications Program.

The Forest Classification and Inventory System (FOCIS) used machine-assisted processing techniques to extract and process tonal, textural, and terrain information from registered Landsat MSS data and digital terrain data. Employing these techniques for processing imagery and field scientific data for the western part of the Klamath National Forest (an area of 390,000 hectares), an estimate of softwood timber volume was obtained with a coefficient of variation of 6.3 percent. This estimate is quite similar to accuracies derived concurrently by Forest Service personnel using conventional techniques, yet was derived in considerably less time and at a lower cost than the estimates produced by the conventional survey (Franklin et al., 1986; Strahler et al., 1981; Woodcock et al., 1982).

Other forestry projects demonstrated strong correlations between stand density and Landsat remote sensing images for single species plantations, including stands of ponderosa pine, southern pine forests, and stands of Douglas fir and red and white fir (Lawrence and Herzog, 1975; Williams, 1976; Strahler et al., 1978). Other efforts at computer-assisted classification of forest vegetation types have been reported for a variety of geographic regions: Bryant et al. (1980), for example, used machine-assisted

processing of Landsat data from northern Maine to produce classified for-
est-type maps to within 95 percent of conventional inventory figures.
Green (1981) produced a Landsat classification map of Bangladesh forests.
In this work, he reports that forest cover has decreased by 85 percent since
1900. Miller and Williams (1979) note that exploitation of Nigerian forests
can be mapped using Landsat data, while Miller et al. (1979) report on
shifting cultivation in forests of northern Thailand using Landsat. Joyce et
al. (1980) employed machine-assisted processing of Landsat data to analyze forest land use change in Louisiana.

If we examine recent trends in the literature of spaceborne remote
sensing for forest classification (Klankamsorm, 1978; Sugarbaker et al.,
1980; Gregg et al., 1979; Malila, 1980; Kirkpatrick and Dickinson, 1982;
Dodge and Bryant, 1978; Bryant et al., 1980), we find that classification
accuracies typically run between 43 percent and 98 percent with a mean
of 72 percent for 3 to 11 classes. Accuracies tend to improve and a number
of the accuracies reported earlier could possibly have been improved if a
geographic or spatially referenced information systems approach to classification had been employed. Even so, based on the results of these studies, it appears that the Landsat sensor system is able to provide systematic,
repetitive, mesoscale data of the Earth's surface of adequate quality to
evaluate the spatial distribution of major forest types of the world.

Global Vegetation Monitoring

Finally, a growing number of authors have recently advocated the use
of meteorological satellites, such as the polar-orbiting Advanced Very
High Resolution Radiometer (AVHRR), for vegetation monitoring (Norwine and Greegor, 1983; Tucker et al., 1985). The AVHRR on the NOAA-
6 and NOAA-7 satellites have an IFOV of 1.1 km at nadir and a 110.8
degree scan angle. These systems image surface cover on a daily basis.
AVHRR data are being used by USDA and NASA researchers to facilitate
large-area estimations of vegetation cover types and/or conditions, and to
assess vegetation phenology (Justice et al., 1986), detect stress in cultural
vegetation (Duggin, 1983), and assess deforestation in tropical forests
(Tucker et al., 1983).

Results of recent research on the use of AVHRR for monitoring tropical
deforestation in Rondonia, Brazil (Tucker et al., 1983) suggest that
AVHRR channel 3 (3.5 μm–3.9 μm) appears to be more sensitive to forest
disturbance than channel 1 (0.55 μm–0.68 μm) and channel 2 (0.73 μm–
1.1 μm). Transformations of AVHRR channels 1 and 2 appear to correlate
well to results achieved employing Landsat channel transforms, but also

appear to have the same sensitivity to phase angle changes as Landsat data. The large scan width (area coverage) and the attendant problems of geometric and atmospheric distortions associated with AVHRR data could limit the potential of channel transforms for vegetation survey, unless images from near-nadir viewing geometries are used or multiview-angle images are radiometrically synchronized. Despite these problems induced by large AVHRR view angles, the potential for daily coverage of surface targets is important, particularly in areas of high cloud cover.

Because AVHRR has finer temporal resolution (compared to the Landsat MSS) as well as adequate spectral resolution for sensing vegetation, the potential of AVHRR for large-scale ecosystems analysis should definitely be pursued in detail. For AVHRR data to be useful in global vegetation surveys, however, further research is still required to address problems of data collection, "ground-truth" assessment, and adjustments to handle pixel size. What is needed is: 1) ground data collection that verifies the accuracy of the remotely sensed data and permits extrapolation of biophysical parameters. Sample areas would have to be two to three times AVHRR pixel size (Justice and Townsend, 1981); 2) assessment of biological relevance. Whether or not spectral classes resolvable by the AVHRR are useful for classifying and inventorying important plant ecosystems must be determined; and 3) methods for minimizing image geometric and radiometric variations produced by changes in pixel size and recording conditions. Despite these gaps in our knowledge, it appears from recent results (Logan, 1983) that the AVHRR can play a significant role in global vegetation assessments in the future.

Biophysical Remote Sensing

A more general research advance of the last decade has produced a better understanding of how different wavelength bands of the electromagnetic spectrum, imaged by aircraft and satellite sensor systems, provide different kinds of information concerning vegetation types, physiological status, and biophysical characteristics (Curran, 1980; Jensen, 1983). In this area, researchers have demonstrated how the ratio of different spectral bands yields information not obtainable directly from single bands spectral information alone, that the near-infrared spectral region (0.7 μm–1.1 μm) exhibits some sensitivity to total plant biomass (Tucker, 1978). Researchers have also demonstrated that healthy green vegetation is typically characterized by high reflectance (45% to 50%), high transmittance (45% to 50%), and low absorbance of spectral radiation from 0.7 μm–1.1 μm (Sinclair et al., 1971). The analyses conducted by these researchers have generally involved the use of transforms of the multispectral image data of the areas under investigation. Most of these "indices," or transformations, use ratios of measurements taken from at least one band in the near-infrared region (0.7 μm–0.9 μm) and one band in the red region (0.6 μm–0.7 μm) of the electromagnetic spectrum. A linear combination of the bands has also been shown to be more highly correlated with biomass than either red or near-infrared measurement alone (Tucker, 1979; Curran, 1980; Bartlett and Klemas, 1980). Examples of the types of transforms that may be applied to both Landsat Thematic Mapper and Multi-spectral scanner data sets are seen in Table IV–5.

Kauth and Thomas (1976), employing multispectral scanner data, demonstrated that most of the spectral variability was two-dimensional. These researchers developed a linear orthogonal transformation with one axis representing soil brightness and the other representing a measure of the vegetation development. This second axis, sensitive to vegetation, was termed *greenness*. It was also shown that this greenness transform is insensitive to both shadow effects and atmospheric effects within a reasonable range of atmospheric conditions. The Kauth-Thomas transformation has also been shown to minimize differences due to soil types and soil moisture conditions (Kollenkark et al., 1982). Thus employing this transform for geographic areas with reasonably limited variations in soil greenness (typed numerical value of soil) can be assumed a reliable surrogate for vegetation development. These characteristics make the greenness index an attractive variable to explore for determining leaf area index and biomass. Indeed, a number of researchers have been exploring the potential of remote sensing techniques for estimating these biophysical characteristics of natural vegetation (Botkin et al., 1980, 1982; Woodwell et al., 1983). The potential of remote sensing to provide statistically significant correlations between reflectance and LAI, or biomass with r^2 typically between 5 and 8 for LAI up to 7 or 8, have been confirmed (Rouse et al., 1973; Wiegand and Richardson in Botkin et al., 1980). Most recently, researchers from the University of California at Santa Barbara have been attempting to bring together the results of work conducted on biophysical analysis of vegetation characteristics with the work on large area inventory surveying and mapping. Using an information systems approach, these researchers are hoping to demonstrate the degree to which satellite aircraft and surface sampling can be combined to produce regional, continental, and global estimates of important biogeochemical, biophysical, and ecological aspects of the surface cover of the Earth. The direction this research is taking is discussed in more detail in the next section.

TABLE IV-5 SOME TRANSFORMS APPLIED TO MSS AND TM DATA TO ENHANCE SELECTED CHARACTERISTICS OF VEGETATION

1. MSS7/MSS5	Ratio (R)
2. MSS6/MSS5	Ratio (R)
3. (MSS6 − MSS5)/(MSS6 + MSS5) (Vegetation Index Using MSS6 "VI6"	Ratio (R)
4. (MSS7 − MSS5)/(MSS7 + MSS5) VI7	Ratio (R)
5. SQRT(VI6 + 0.5) Transformed Vegetation Index Using VI6 "TVI6"	Ratio (R)
6. SQRT(VI7 + 0.5) TVI7	Ratio (R)
7. Local Standard-Deviation (SD) Texture of MSSS7/MSS5	Ratex (RT)
8. Local-Mean (M) Texture of MSS7/MSS5	Ratex (RT)
9. (MSS7/MSS5)/PC2	Complex (C)
10. (Local SD Texture of MSS7/MSS5)/PC2	Complex (C)
11. (Local SD Texture of MSS7)/(Local SD Texture of MSS5)	Ratex (RT)
12. (MSS7/MSS5)/(Local SD Texture of MSS7/MSS5)	Ratex (RT)

FUTURE PERSPECTIVE

More than a century of development in remote sensing coupled with recent advances in computer science have led us to a new potential for conducting global ecological research (National Research Council, 1986). As seen in Figure IV–4, development of platforms and sensor systems, scales of analysis, data types, analytical techniques, and types of analysis have grown from generally simple systems to much more complex systems.

These advances have led to the tremendous potential of remote sensing today. The would-be user of this technology is faced with a complex array of systems and techniques, each able to produce prodigious amounts of data. However, what one really needs is a straightforward answer to a specific question to improve one's understanding of a particular issue. The procedures used are fundamentally deductive. They employ the basic elements of image analysis and can involve simple interpretation techniques as well as sophisticated multistage sampling strategies, as seen in Figure IV–5. The basic process, however, is to narrow continually the amount of information that has been developed until users have answered their questions. This process is diagrammed in Figure IV–6. Much effort is directed toward this process, yet more needs to be done. These studies will involve research into all areas of the remote sensing information flow. Figure IV–7 presents an idealized version of such a flow.

The keys to future developments are in the areas of information sciences

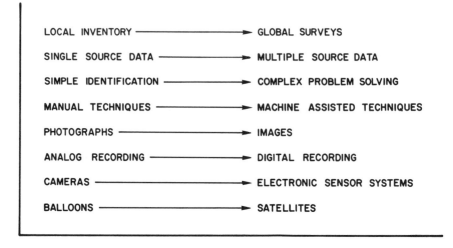

FIGURE IV-4   Development of remote sensing (complexity).

and artificial intelligence. The development of improved analytic techniques through expert system tools and geographic information systems will significantly advance our potential to conduct future global-scale ecological research.

More specifically, in the area of vegetation analysis, remote sensing techniques now appear to offer the capability to estimate total biomass as well as to discriminate green vegetation tissue from woody tissue and to differentiate these from the water contained in plants. In addition, remote sensing appears to offer methods for determining the rates of chemical elements and energy fluxes in ecosystems, to determine rates of primary productivity, and to improve our understanding of biogeochemical cycling. Remote sensing of land surfaces and their vegetation has great practical implications as well. It provides the potential to track the spread of plant diseases and forecast crops, and it can also facilitate the study of global environmental issues. Quantitative, variable estimates of biomass and net primary production (essential to an understanding of the carbon cycle), the carbon dioxide problem, and almost all global ecological issues appear approachable for the first time.

Exploitation, then, of the improved and unique information available to ecologists from remote sensing has barely begun. Yet many scientific and

RELATION OF CLASSIFICATION LEVEL AND SPATIAL RESOLUTION

LEVEL I: Global
AVHRR
resolution: 1.1km

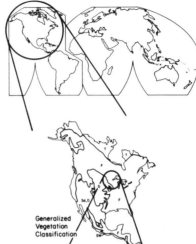

LEVEL II: Continental
AVHRR
Landsat Multispectral Scanner
resolution: 1.1km – 80m

Generalized
Vegetation
Classification

LEVEL III: Biome
Landsat Multispectral Scanner
Thematic Mapper
resolution: 80m – 30m

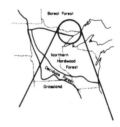

LEVEL IV: Region
Thematic Mapper
High Altitude Aircraft
resolution: 30m – 3m+

Boundary Waters Canoe Area

LEVEL V: Plot
High and Low Altitude Aircraft
resolution: 3m+ – 1m+

Typical Study Area

LEVEL VI: Sample Site
Surface Measurements and
Observations

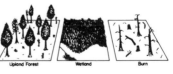

Upland Forest   Wetland   Burn

FIGURE IV–5   Multistage sampling strategies.

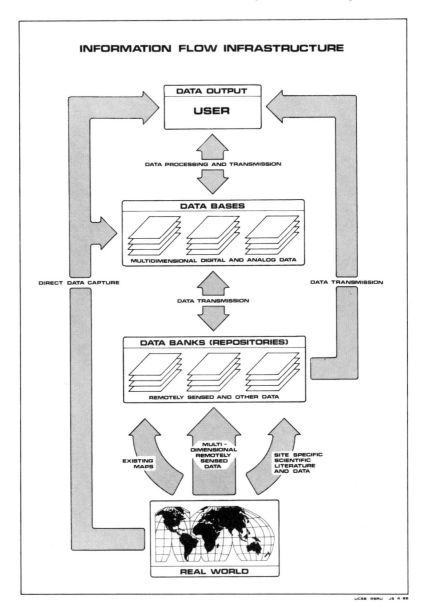

FIGURE IV–6   Relation between real world and user.

IDEALIZED REMOTE SENSING DATA / INFORMATION FLOWS

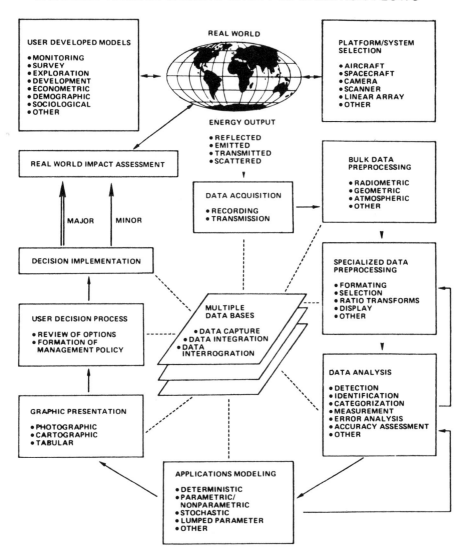

FIGURE IV–7   Idealized remote sensing data/information flow.

technical capabilities exist. Therefore, what is required to increase further the potential of remote sensing for the study of global ecology is a concerted effort on the part of the ecologists to learn the capabilities and limitations of these new techniques. In addition, experts in the remote sensing of the land surface should communicate the capabilities and limitations of the techniques clearly. When analyzed, remote sensing information can provide ecologists with significant improvements in the quantity, quality, and timeliness of the data required for different geographic types and scales of analysis. Remote sensing allows us to expand our information horizons beyond those thought possible only a few years ago. As more ecologists become aware of the implications of remote sensing, the true impact of these techniques on the study of the biosphere, and of biomes and ecosystems, will be felt. When allied with appropriate field studies and ecological modeling, remote sensing can change our perception of the landscape, of the land's biota, and of the interrelationships between life on the land and the rest of the biosphere.

SUMMARY

Remote sensing, especially when combined with ground data, is a useful tool for the study of the biosphere. It can supply information on the terrestrial biota that is crucial to the study of global biophysical processes. From the time of the first photograph, the need for information about the environment has led to rapid development in the remote sensing field. In the area of platforms, we have progressed from nonmaneuverable, tethered balloons to satellites in a variety of orbital configurations. Sensor systems have developed from simple cameras using panchromatic film to array area imaging spectrometers capable of recording over one hundred channels of spectral data on digital tape.

In scale of analysis we have gone from the ability to acquire data of limited local environments to global coverage. In air photo interpretation, often only the air photos were used in analysis. Now it is not uncommon in a remote sensing research project to employ remotely sensed and ancillary or collateral data from a variety of sources (e.g., soil maps, land use maps, literature on species in the area, etc.). Until the early 1960s, manual analysis was the dominant form of investigative technique employed. Now, although manual analysis is still suitable for many applications, computer-assisted analysis is making good progress. Finally, practitioners have changed their approach from a direct measurement orientation to a more complex genre in which the remotely sensed data is but one input to sophisticated modeling procedures. These advances have led to the tremendous potential of remote sensing today.

REFERENCES

Bartlett, D.S. and V. Klemas. 1979. Quantitative assessment of emergent biomass and species composition in ideal wetlands using remote sensing. Workshop on Wetland and Estuarine Processes and Water Quality Modeling, New Orleans, Louisiana, June 10–18, 1979.

Bartlett, D. and V. Klemas. 1980. Quantitative assessment of tidal wetlands using remote sensing. *Environmental Management* 4:337–345.

Bentley, R.G., Jr., B.C. Salmoni Drexler, W.J. Bonner, and R.K. Vincent. 1976. A Landsat study of ephemeral and perennial rangeland vegetation and soils. Final Report, BLM/USDI YA–300-1700-1012.

Bryant, E., A.G. Dodge, Jr., and S.D. Warren. 1980. LANDSAT for practical forest type mapping: A test case. *Photogrammetric Engineering and Remote Sensing* 46(12):1575–1584.

Carneggie, D.M., S.D. DeGloria, and R.N. Colwell. 1975. Usefulness of Landsat data for monitoring plant development and range conditions in California's annual grasslands. *Proceedings NASA Earth Resources Survey Symposium*, Houston, Texas, NASA TMX58168, pp. 19–44.

Carneggie, D.M., B.J. Schumpf, and D.A. Movat. 1983. Rangeland applications. In: *Manual of Remote Sensing*, 2d ed. Vol. 2. American Society of Photogrammetry, Falls Church, Virginia, pp. 2325–2334.

Carter, V. 1976. Applications of remotely sensed data to wetland studies. *Proceedings of 19th COSPAR Meeting*, Philadelphia.

Carter, V., M.K. Garrett, L. Shima, and P. Gammon. 1977. The Great Dismal Swamp: Management of a hydrologic resource with aid of remote sensing. *Bulletin of Water Resources* 13(1):1–12.

Clark, Walter. 1967. Aerial photography by infrared, a few historical notes. *Workshop— Infrared Color Photography in the Plant Sciences*. Florida Department of Agriculture, Division of Plant Industry.

Colwell, R.N. 1956. Determining the prevalence of certain cereal crop diseases by means of aerial photography. *Hilgardia* 26:223–286.

Colwell, R.N., ed. 1960. *Manual of Photographic Interpretation*, Falls Church, Virginia, American Society of Photogrammetry, 868 pp.

Colwell, R.N., ed. 1983. *Manual of Remote Sensing*, 2d ed. American Society of Photogrammetry, Falls Church, Virginia, 2,440 pp.

Curran, P. 1980. Multispectral remote sensing of vegetation amount. *Progress in Physical Geography* 4:315–341.

Dodge, A.G., Jr., and E.S. Bryant. 1976. Forest type mapping with satellite data. *Journal of Forestry* 74(8):526–531.

Duggin, M.J. 1983. The effect of irradiation and reflectance variability on vegetation condition assessment. *International Journal of Remote Sensing* 4(3):601–608.

Ernst, C.L. and R.M. Hoffer. 1979. Digital processing of remotely sensed data for mapping wetland communities. LARS Technical Report 122079, Indiana Purdue University, West Lafayette.

Ernst, C.L. and R.M. Hoffer. 1981. Using Landsat MSS data with soils information to identify wetlands habitats. *Satellite Hydrology V Pecora Memorial Symposium*, Sioux Falls, South Dakota, pp. 474–478.

Estes, J. 1969. *Geographic Applications of Multi-Image Correlation Remote Sensing Techniques.* University Microfilms, Inc., Ann Arbor, Michigan, 227 pp.

Estes, J.E., J.R. Jensen, and D.S. Simonett. 1980. Impacts of remote sensing on U.S. geography. *Remote Sensing of Environment* **10**:43–80.

Estes, J.E. and D.S. Simonett. 1975. Fundamentals of image interpretation. In: *Manual of Remote Sensing*, J.E. Estes and D.S. Simonett, eds. American Society of Photogrammetry, Falls Church, Virginia, pp. 869–1076.

Fisher, W.A., ed. 1975. History of remote sensing. *Manual of Remote Sensing*, R.G. Reeves, ed. American Society of Photogrammetry, Falls Church, Virginia, pp. 27–50.

Gammon, P.T. and V.P. Carter. 1976. Comparison of vegetation classes in the Great Dismal Swamp using two individual Landsat images and a temporal composite. *Proceedings of Purdue Symposium on Machine Processing of Remotely Sensed Data*, p. 3B–1.

Goward, S.N., C.J. Tucker, and D.G. Dye. 1985. North American vegetation patterns observed with the NOAA–7 Advanced Very High Resolution radiometer. *Vegetatio* **64**:3–14.

Green, K.M. 1981. Digital processing of forest habitat in Bangledesh and the development of a low-cost processing facility at the National Zoo, Smithsonian Institution. *Proceedings, 15th International Symposium on Remote Sensing of Environment*, vol. 13, Environmental Research Institute of Michigan, Ann Arbor, Michigan, pp. 1315–1325.

Gregg, T., E. Barthmaier, R. Aulds, and R. Scott. 1979. *LANDSAT Operational Inventory Study.* State of Washington, Department of Natural Resources, Olympia, Washington, 18 pp.

Haas, R.H., D.W. Deering, J.W. Rouse, Jr., and J.A. Schell. 1975. NASA Earth Resources Survey Symposium, Houston, Texas, pp. 43–52.

Harman, W.E., Jr., ed. 1966. Aerial photography. *Manual of Photogrammetry*, 3d ed. American Society of Photogrammetry, Falls Church, Virginia, Vol. 1, pp. 195–242.

Heller, R.C. and J.J. Ulliman. 1983. Forest resource assessments. In: *Manual of Remote Sensing*, 2d ed., Vol. 2, American Society of Photogrammetry, Falls Church, Virginia, pp. 2229–2324.

Howard, J.A. 1970. *Aerial Photo-Ecology.* American Elsevier Publishing Company, New York, 325 pp.

Jensen, J.R. 1980. Biophysical remote sensing. *Annals of the Association of American Geographers* **73**:111–132.

Johnston, A. 1978. Remote sensing for estimation of rangeland productivity and condition. *Research Highlights*, Lethbridge, Alberta, Canada.

Joyce, A.T., J.H. Ivey, and G.S. Burns. 1980. The use of Landsat MSS data for detecting land use changes in forestland. *Proceedings 14th International Symposium on Remote Sensing of Environment*, Vol. 2, Environmental Research Institute of Michigan, Ann Arbor, Michigan, pp. 979–988.

Justice, C.O., J.R.G. Townshend, B.N. Holben, and C.J. Tucker. 1986. Analysis of the phenology of global vegetation using meteorological satellite data. *International Journal of Remote Sensing* **6**:1271–1318.

Kan, E.P. and F.P. Weber. 1978. The Ten-Ecosystem Study: Landsat mapping of forest and rangeland in the United States. *Proceedings, 12th International Symposium on Remote Sensing of Environment*, Vol. 3, Environmental Research Institute of Michigan, Ann Arbor, Michigan, pp. 1809–1825.

Kauth, R.G. and G.S. Thomas. 1976. The tasseled cap, a graphic description of the spectral-temporal development of agricultural crops as seen by Landsat. *Proceedings, Symposium on Machine Processing of Remotely Sensed Data*, LARS Purdue, IEEE Cat. No. 76, 6.23–7.2.

Kemp, R.C., C.G. Lewis, C.W. Scott, and C.R. Robinson. 1925. Aero Photo Survey and Mapping of the Forests of the Irrawaddy Delta, Burma Forest Bulletin, No. 11, 42 pp.

Kirkpatrick, J.B. and K.J.M. Dickinson. 1982. Recent destruction of natural vegetation in Tasmania. *Search* **13**:186–187.

Klankamsorn, B. 1978. Use of satellite imagery to assess forest deterioration in eastern Thailand. *Proceedings, 12th International Symposium on Remote Sensing Environment*, Environmental Research Institute of Michigan, Ann Arbor, Michigan, pp. 1299–1306.

Klemas, V., D. Bartlett, and R. Rogers. 1975. Coastal zone classification from satellite imagery. *Photogrammetric Engineering and Remote Sensing* **41**:533–542.

Kohler, R.J. and H.K. Howell. 1963. Photographic enhancement by superimposition of multiple images. *Photographic Science and Engineering* **7**:241–245.

Kollenkark, J.C., C.S.T. Daughtry, M.E. Bauer, and T.L. Hausley. 1982. Effects of cultural practices on agronomic and reflectance characteristics of soybean canopies. *Journal of Agronomy* **74**:751–758.

Kuchler, A.W. 1973. Problems in classifying and mapping vegetation for ecological regionalization. *Ecology* **54**:512–523.

Land, E.H. 1959. Color vision and the natural image, part 1. *National Academy of Sciences, Proceedings* **45**:115–129.

Lawrence, R.D. and J.H. Herzog. 1975. Geology and forestry classification from ERTS–1 digital data. *Photogrammetric Engineering and Remote Sensing* **41**:1241–1251.

Lent, J.D. and G.A. Thorley. 1969. Some observations on the use of multiband spectral reconnaissance for the inventory of wildland resources. *Remote Sensing of Environment* **1**:31–45.

Lo, C.P. 1976. *Geographical Applications of Aerial Photography*. Crane, Russak and Company, New York, 330 pp.

Logan, T.L. 1983. *Regional Biomass Estimation of a Coniferous Forest Environment from NOAA-AVHRR Satellite Imagery*, University of California, Santa Barbara, Doctoral Dissertation, 263 pp.

Lulla, K. 1983. The Landsat satellites and selected aspects of physical geography. *Progress in Physical Geography* **7**:1–45.

MacDonald, R.B., M.E. Bauer, R.D. Allen, J.W. Clifton, J.D. Erickson, and D.A. Landgrebe. 1972. Results of the 1971 Corn Blight Watch Experiment. *Proceedings of the 8th International Symposium on Remote Sensing and Environment*, University of Michigan, pp. 157–185.

MacDonald, R.B. and F.G. Hall. 1980. Global crop forecasting. *Science* **208**:670–679.

Malila, W.A. 1980. Change vector analysis: An approach for detecting forest changes with LANDSAT. *Proceedings of 6th International Symposium on Machine Processing of Remotely Sensed Data*, Purdue University, West Lafayette, Indiana, pp. 326–336.

Mazade, A.V., ed. 1981. The Ten-Ecosystems Study, Final Report. LEMSCO–13491. Lockheed Engineering and Management Service Co., Houston, Texas.

McCamy, C.S. 1960. A demonstration of color perception and abridged color-projection systems. *Photographic Science and Engineering* **4**:115–159.

Miller, L.D., K. Nualchawee, and C. Tom. 1978. *Analysis of the Dynamics of Shifting Cultivation in the Tropical Forests of Northern Thailand Using Landscape Modelling and Classification of LANDSAT Imagery*, NASA Technical Memorandum No. 79545, Greenbelt, Maryland.

Miller, L. and D. Williams. 1979. Forest exploitation in the Central Nigerian forest reserves. In: *Monitoring Forest Canopy Alteration Around the World with Digital Analysis of LANDSAT Imagery*, NASA, Goddard Space Flight Center, Greenbelt, Maryland, pp. 9–13.

Morrisson, R.B. and M.E. Cooley. 1973. Application of ERTS-1 multispectral imagery to monitoring the present episode of accelerated erosion in southern Arizona. *Symposium on Significant Results*, SP327 NASA Document, Washington, DC, pp. 283–290.

Myers, V.I., ed. 1983. Remote sensing applications in agriculture. In: *Manual of Remote Sensing*, 2d ed., Chapter 33, *American Society of Photogrammetry*, Falls Church, Virginia, pp. 2111–2220.

National Research Council. 1986. *Remote Sensing of the Biosphere*. Washington, DC, National Academy Press, 135 pp., NASA, 1983. *Land-related Global Habitability Science Issues*, NASA Technical Memorandum 85841, National Aeronautics and Space Administration, Scientific and Technical Information Branch, 112 pp.

Norwine, J. and D.H. Greegor. 1983. Vegetation classification based on Advanced Very High Resolution (AVHRR) satellite imagery. *Remote Sensing of Environment* 13:69–87.

Powers, F.G. and C. Gengry. 1979. *Operation Overflight*. Holt, Rinehart and Winston, New York, 375 pp.

Quackenbush, R.S., Jr. 1960. Development of photo interpretation. In: *Manual of Photographic Interpretation*, American Society of Photogrammetry, Falls Church, Virginia, pp. 1–18.

Reeves, R.G., ed. 1975. *Manual of Remote Sensing*. Falls Church, Virginia, American Society of Photogrammetry, 2144 pp.

Rouse, J.W., Jr., R.H. Haas, J.A. Schell, and D.W. Dearing. 1973. Monitoring vegetation systems in the Great Plains with ERTS. In: *NASA Goddard Space Flight Center, Earth Resources Technology Satellite-1 Symposium*. Washington, DC, December 1983, Proceedings, Vol. 1, Section A, pp. 309–319.

Shasby, M.B., R.R. Burgan, and G.R. Johnson. 1981. Broad area forest fuel and topography mapping using digital LANDSAT and terrain data. *Proceedings, 1981 Machine Processing of Remotely-Sensed Data Symposium*, pp. 529–538.

Sinclair, R.T., R.M. Hoffer, and M.M. Schreiber. 1971. Reflectance and internal structure of leaves from several crops during a growing season. *Agronomy Journal* 63:864–868.

Spurr, H. 1960. *Photogrammetry and Photo-Interpretation*, The Ronald Press Company, New York, 472 pp.

Strahler, A.H. 1978. Improving forest cover classification accuracy from Landsat by incorporating topographic information. Twelfth International Symposium on Remote Sensing of Environment. Manila, Philippines, pp. 927–942.

Strahler, A.H., J. Franklin, C.E. Woodcock, and T.T. Logan. 1981. FOCIS: A forest classification and inventory system using Landsat and digital terrain data. Final Report, NASA Contract No. NAS–915509, University of California, Santa Barbara, 60 pp.

Strahler, A.H., T.L. Logan and N.A. Bryant. 1978. Improving forest cover classification accuracy from Landsat by incorporating topographic information. *Proceedings of the 12th International Symposium on Remote Sensing of Environment*, Manila, The Philippines, Environmental Research Institute of Michigan.

Sugarbaker, L.T., E. Gregg, E. Barthmaier, R. Scott, M. Fleming, and R. Hoffer. 1980. *Forest Cover Type Mapping in Eastern Washington with LANDSAT and Digital Terrain Data*, Pacific Northwest Regional Commission Grant No. 1080613.

Suits, G.H. 1972. The calculation of the directional reflectance of a vegetative canopy. *Remote Sensing of Environment* 2:117–125.

Swain, P.H. 1979. Pattern recognition. In: *Quantitative Remote Sensing*, Swain and Davis, eds. McGraw-Hill, New York, pp. 35–60.

Thomas, H.H. 1920. Aircraft photography in the service of science. *Nature* 105:257–259.

Tucker, C. 1978. A comparison of satellite sensor bands for vegetative monitoring. *Photogrammetric Engineering and Remote Sensing* 44:1369–1380.

Tucker, C. 1979. Red and photographic infrared linear combinations for monitoring vegetation. *Remote Sensing of Environment* **18**:127–150.

Tucker, C.J. and M.W. Garratt. 1977. Leaf optical system modeled as a stochastic process.
Applied Optics **16**:635–642.

Tucker, C.J., J.A. Gatlin, and S.R. Schneider. 1984. Monitoring vegetation in the Nile delta
with NOAA–6 and NOAA–7 AVHRR imagery.

Tucker, C.J., J.R.G. Townshend, and T.E. Goff. 1985. African landcover classification satellite
data. *Science* **227**:369–375. *Photogrammetric Engineering and Remote Sensing* **50**:53–61.

Williams, D.L. 1976. A canopy-related stratification of a southern pine forest using Landsat
digital data. *Proceedings of the 1976 Fall Convention of the American Society of Photogrammetry,
Seattle, Washington.* American Society of Photogrammetry, Falls Church, Virginia.

Woodcock, C.E., J. Franklin, A.H. Strahler, and T.C. Logan. 1982. Improvements in forest
classification and inventory using remotely sensed data. *Proceedings of the 16th International
Symposium on Remote Sensing of Environment.* Environmental Research Institute of Michigan,
Ann Arbor, pp. 563–574.

V BIOGEOCHEMICAL CYCLES

B. MOORE III, M. PATRICIA GILDEA,
C.J. VOROSMARTY, D.L. SKOLE, J.M. MELILLO,
B.J. PETERSON, E.B. RASTETTER
AND P.A. STEUDLER

INTRODUCTION

The recognition of global ecology as a legitimate and imperative discipline is a direct result of the increasingly complex effects of human global habitation. The study of global biogeochemical cycles is essential to global ecology (Moore et al., 1984). With hydrogen and oxygen, four other elements—carbon, nitrogen, phosphorus, and sulfur—are of special interest in the study of our planet. Because of life, each of these four elements follows a closed loop or cycle through increasing molecular energy states as the elements are incorporated into living cells of organisms and then decreasing energy levels as the organisms decompose. These cycles are an expression of life; they are the metabolic system of the planet. Their various patterns are the consequence of myriad biological, chemical, and physical processes that operate across a wide spectrum of time scales. In the absence of significant disturbance, these processes define a natural cycle for each element with approximate balances in their sources and sinks, which result in a quasi-steady state for the cycle, at least on time scales less than a millennium.

Human activity since the Industrial Revolution has increased to such an extent that it must now be regarded as a significant perturbation to these critical biogeochemical cycles of our planet. The magnitude of human activity is global and the effects are growing. Certain indicators of the state of particular cycles, such as levels of atmospheric carbon dioxide, nitrous oxide, carbon monoxide, and methane for the carbon cycle, have moved well outside the recent historical distributions (Bacastow and Keeling, 1981; Khalil and Rassmussen, 1983a; Khalil and Rassmussen, 1983b; Khalil and Rassmussen, 1984; Banin et al., 1984). Similarly, changes in the nitrogen and sulfur cycles are reflected by the onset of what we now call acid rain. It is difficult to identify a major river or estuary that has not been affected by the addition of phosphate from agricultural, urban, or industrial sources (Moore, 1985; Peterson, 1981; Steudler and Peterson, 1985).

The biological and chemical webs of elements critical to Earth's biota integrate traditional ecological divisions of terrestrial, aquatic, coastal, and open ocean systems of the Earth's surface. The study of these cycles has at least two fundamental goals. First is the characterization of the natural elemental cycles and the linkages among the four elements significant to Earth biota, namely, carbon, nitrogen, sulfur, and phosphorus. Second is the identification and quantification of changes in these cycles due to anthropogenic activity.

This chapter describes the use of mathematical models as a development strategy for representing the global complexities of carbon, nitrogen, phosphorus, and sulfur cycles. While we realize that a complete and detailed representation of these complexities is still not yet an achievable end, an initial theoretical framework has been established. An approach has been defined to identify and quantify the ecological phenomena critical to the functioning of the Earth's biosphere.

Only with the current technological advances in remote sensing and computer systems have explorations of ecological phenomena been possible at the global scale. Evolution of imagery techniques created the ability to gather the enormous amount of information necessary for quantification of aspects of the Earth's surface. Development of computer systems provided the facility to store, retrieve, and manipulate these global data bases. These interdisciplinary data bases require a framework or model to describe complex relationships in global nutrient cycles.

These relationships, their complexity, and their scope are relatively new ecological concerns. A question central to this development strategy is whether the appropriate paradigms exist to support models of global-scale ecological interactions. We believe that useful approaches exist for studying units of landscape at global resolution by considering energy flows and element fluxes within and between adjacent ecosystems.

For example, at the small watershed level we understand how the energetics of a stream draining a watershed depends on the quantity and quality of the reduced carbon compounds entering the stream from the upland (Likens et al., 1977; Johnson and Swank, 1973). We have also learned that a disturbance in the upland not only can change the quality and quantity of nutrients (N, P) flowing from the upland to connected aquatic systems, but also can affect nutrient exchange (via gases and particulates) with the atmosphere. Our challenge is to devise a protocol for scaling up the small watershed approach to larger units of landscape and the linkages among them.

IMPACTS OF HUMAN ACTIVITY

The critical role of living systems in the Earth's geochemical cycles is a relatively recent discovery. The realization that biotic factors may provide homeostatic controls helps us recognize the natural metabolism responsible for the Earth's atmosphere, ocean, and sediment. Planetary metabolism is interactive; clearly quantification of the contribution of the biota is essential for a better understanding of global processes (see chapter by Lovelock and Margulis, this volume). The restricted elemental composition of biological systems limits the possible chemical transformations of hydrogen, oxygen, nitrogen, phosphorus, and sulfur. However, our knowledge of the way these biogeochemical cycles relate to each other compares poorly with our general understanding of the individual cycles (Peterson and Melillo, 1985).

Human activity has apparently significantly altered several elements. Some of these changes are well documented on the global scale. For example, estimates concerning the production of CO_2 from fossil fuel combustion are now widely accepted (Rotty, 1981). Although tentative estimates of the changes in global stocks of carbon in terrestrial ecosystems have improved (Moore et al., 1981; Houghton et al., 1983; Houghton et al., 1985), the consequences of changes in the carbon cycle for the cycles of nitrogen, phosphorus, and sulfur are just now being evaluated (Gildea et al., 1986; Vorosmarty et al., 1986).

Major questions remain regarding the details of individual cycles. Although CO_2 has been accumulating in the atmosphere for at least the last 20 years at the rate of 1 ppmv per year, a balance between sources and sinks of atmospheric CO_2 is still elusive (Moore and Bolin, 1987). Simply stated, the annual budget (Table V–1) does not balance unless:

1. Fertilization effects, either terrestrial or aquatic, equal deforestation-regrowth;

TABLE V-1   IMBALANCES IN THE GLOBAL CARBON BUDGET: AN EXCESS OF
CARBON SOURCES OVER SINKS

Since atomospheric increase and fossil fuel loading are well-established, the budget can be
balanced only if the net deforestation term is diminished, the fertilization estimate (both
aquatic and terrestrial) is increased, and/or the oceanic uptake is underestimated. There
may also be natural changes in biotic uptake that are not now elucidated. The fluxes below
represent net carbon movement for each category of sources and sinks.

	Input into Atmosphere
Fossil Fuel	$5 \times 10^5 \mathrm{gC\ y^{-1}}$
Deforestation-Regrowth	$2 \times 10^{15} \mathrm{gC\ y^{-1}}$
	$7 \times 10^{15} \mathrm{gC\ y^{-1}}$
	Uptake
Atmospheric Increase	$2.5 \times 10^{15} \mathrm{gC\ y^{-1}}$
Oceanic Uptake	$2.5 \times 10^{15} \mathrm{gC\ y^{-1}}$
Fertilization Effects	?
	$5 \times 10^{15} \mathrm{gC\ y^{-1}} + ?$

2. The problem vanishes (i.e., deforestation equals regrowth); or
3. The oceanic uptake of carbon dioxide is grossly underestimated.

To appreciate the constraints upon these possibilities, the basic global carbon budget (Figure V–1) is reviewed briefly.

The present concentration of CO_2 in the atmosphere is 340 ppmv or, by
mass of carbon (C), $728 \times 10^{15} \mathrm{gC}$. In a single year, seasonal differences
in photosynthesis and respiration within the biosphere create an oscillation
in the atmospheric CO_2 concentration with an amplitude of roughly 5
ppmv or $10.6 \times 10^{15} \mathrm{gC\ y^{-1}}$. A few attempts have been made to detect
changes in the amplitude of this oscillation as a way to infer changes in the
activity (photosynthesis, respiration) of biospheric pools of carbon, but the
results have been inconclusive. Estimates of the amount of carbon in living
organic matter on land vary between 450 to $900 \times 10^{15} \mathrm{gC}$. Estimates of
total primary production, respiration, and detrital decay rates also vary
greatly. There are two principal reasons for this uncertainty: 1) the method
of scaling from selected local sites of measurement to biomewide estimates
is not rigorous and 2) uncertainties exist about geographic extent of different biomes.

Although soil organic matter (humus) is a major active reservoir in the
global carbon cycle, there are few data on its size and turnover rate. Esti-
mates range from 700 to $1800 \times 10^{15} \mathrm{gC}$. The total $(1200 \times 10^{15} \mathrm{gC})$

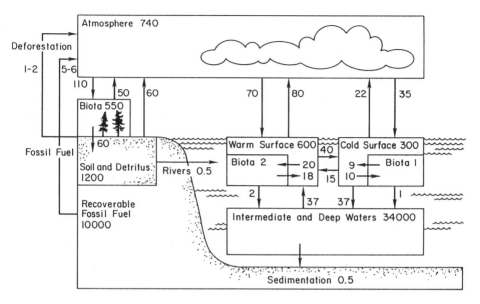

FIGURE V–1   Fundamental features of the global carbon cycle.

shown in Figure V–1 includes human disturbance and revised estimates on the geographical range of various systems. It excludes the slower turnover pools of carbon in deep humus and peat.

The magnitude of the net release of carbon from vegetation and soils of the world as a result of land use changes remains unknown. The problem of calculating this release centers on two general, but important, factors: the rate of land use change and the response of the biota to disturbances. Estimates of the current rate of conversion of closed canopy tropical forests to agriculture vary from $3.5 \times 10^4 \, \text{km}^2 \, \text{y}$ to $15 \times 10^4 \, \text{km}^2 \, \text{y}^{-1}$. Analysis of the biotic response to such disturbance has been equally difficult; improved description of biotic responses requires better measurements of net ecosystem production and/or in situ carbon stocks following disturbance.

Direct calculations of net carbon released from the biosphere have accounted for various estimates of disturbances and the uncertainties of biotic response. From 1860 to 1980, the mean release totaled $150 \times 10^{15} \text{gC}$, with a current annual release between 1 to $3 \times 10^{15} \text{gC} \, \text{y}^{-1}$ (Houghton et al., 1983; Houghton et al., 1985). However, without far more refined information about disturbance rates and ecosystem responses, it will be impossible to appraise with confidence the role of the biota in the global carbon cycle.

Similar uncertainties exist about anthropogenic impacts on the nitrogen cycle. When fossil fuel is burned, not only are large amounts of carbon released to the atmosphere, but there may also be an increase in the input of nitrogen to the atmosphere. Some of this nitrogen may enter terrestrial ecosystems in bulk precipitation. This rise in available nitrogen may stimulate both carbon fixation and storage. Conversely, wood harvest can reduce not only the carbon stock for forest ecosystems, but the nitrogen stock as well.

Humans are modifying the nitrogen cycle in additional ways. First, the cultivation of legumes and the production of artificial fertilizers places more than half the global N_2 fixation process within human influence. Second, industrial agriculture has increased the rate of decomposition of organic matter in the soil and, thus, the return of nitrogen to the atmosphere. Third, high temperature combustion represents 10 to 20 percent of the total flux of fixed nitrogen to the atmosphere and is highly concentrated in the inhabited regions of the Northern Hemisphere. The combustion-produced nitrogen oxides have an important impact on atmospheric chemistry, biological productivity, and the acidity of precipitation in the northern mid-latitudes. Fourth, the concentration of human populations in coastal zones and the routine discharge of wastes and runoff are dramatically affecting coastal aquatic ecosystems.

These sources of accelerated nitrogen cycling all potentially impact on the production of trace gases, such as N_2O, through the coupling of N_2O production with the biological processes of nitrification and denitrification.

Important issues remain unresolved with respect to nitrogen cycling in ecosystems. On land, a major dilemma exists as to why, when nitrogen is abundant in the atmosphere and soils, nitrogen scarcity is generally the element that limits the growth of plants in both natural and agricultural ecosystems. Insights into the recalcitrance of bound nitrogen in the soil may be gained by understanding the relationships and linkages between carbon and nitrogen cycling.

Whether a steady state concentration even exists for oceanic fixed nitrogen (on time scales of 10^4 years) is unclear. The meager existing data suggest imbalance between the sources of oceanic nitrogen (transport from land, input from the atmosphere, and in situ fixation) and ocean losses (denitrification and removal to sediments). If the present estimates of sources and sinks were accurate, oceanic nitrogen should undergo a gradual depletion over a period of about 10^5 years, followed perhaps by major influx of nitrogen into the oceans during any future ice age. This problem is of profound importance, especially from a climatic point of view, because it is believed that both nitrogen and phosphorus could limit the fixation of carbon in the oceans, which in turn could profoundly affect atmospheric carbon dioxide.

Phosphorus also is essential to growth in terrestrial ecosystems and yet, unlike carbon, it is frequently in short supply. Its relative insolubility limits its availability to organisms in soils, rivers, and oceans, and sedimentary deposits provide its major reservoirs. Because it is not volatile, phosphorus plays little role in atmospheric chemistry. In the absence of human activity, these characteristics limit this element's role in global biogeochemical cycles. Human activity, however, has altered the availability of phosphorus in direct and indirect ways. The application of phosphorus fertilizer is a direct perturbation, but subtler alterations of the phosphorus cycle may influence the dynamics of other cycles. Fire, either natural or as a management technique, may increase the available stocks of phosphorus, because oxidation of plant litter transforms organically bound phosphorus into more available forms. Increased levels of available phosphorus can, in turn, raise the rate of restoration of nitrogen to soils.

Sulfur plays a vital role in maintaining biological systems, because it is an essential nutrient for all bacteria, protists, fungi, plants, and animals. Sulfur is a limiting nutrient in parts of Canada's grain belt. In addition, because it is so widely available in seawater (as sulfate), it has a great impact on the cycles of carbon and energy in marine ecosystems. Current understanding of the global sulfur cycle is limited. Most of the reduced sulfur gases are believed to be biogenic in origin, but little data exist regarding the nature or strengths of the sources of their spatial and temporal distributions (Steudler and Peterson, 1984; Sagan, 1985). Furthermore, such reduced gases are oxidized to SO_2 through processes whose mechanisms and rates remain poorly understood. As much as half of the oxidized sulfur gases could be accounted for by such oxidation reactions, the other half derived from fossil fuels. Anthropogenic sulfur emissions may be comparable to release from natural systems. This is in contrast to human contributions to the global cycles of carbon, nitrogen, and phosphorus, where human activity represents relatively minor components of their respective natural fluxes.

OVERVIEW OF MODEL STRUCTURE

In the rest of this chapter we trace the development of incipient models for the description of biogeochemical cycles that approach the complex issues raised earlier. We present here an overview of our models and their evolution. For detailed analysis and description, see Gildea et al. (1986) and Vorosmarty et al. (1986). Model development strategy focuses on two central concerns: the underlying interactions of the cycles and the perturbation of the cycles by human activity. These goals define both the spatial and temporal resolution that such models use. The time span is decadal, encompassing a century or two rather than tens or hundreds of centuries, and incorporates a history of land use changes, such as over the years 1800 to 2000. The models are geographically detailed both for spatial and temporal changes in land use and for spatial variations in ecosystem type and response. They link four principal systems through which nutrients and carbon pass: terrestrial, aquatic and near-shore marine, open ocean, and atmospheric. Finally, they provide an organizing approach for describing the linkages among the systems that integrate the energy and matter transfers in manageable units, but still provide sufficient spatial resolution.

Two possible approaches for organizing system linkages are by watershed or by airshed. Defining ecosystem responses using airsheds as a basic unit integrates terrestrial/aquatic fluxes and provides an appropriate scale for transferring, or cycling, material globally. However, with the inclusion of the critical but nonvolatile element phosphorus, the terrestrial/aquatic interactions need more specific treatment. Using watersheds as an organizational unit couples terrestrial nutrient response with aquatic and riverine nutrient processing in well-bounded units of landscape. Within a watershed (Figure V–2), terrestrial and aquatic systems interact with an atmospheric component, thereby establishing point sources for gas fluxes and possible later linkages to global models of general circulation. The terrestrially derived nutrients are processed in turn by riverine and coastal ocean ecosystems and become the input functions to an open-ocean model, which is adapted from the work of Bolin (1981a, 1981b; Bolin et al., 1983; Moore and Bjorkstrom, 1986). The ocean model considers advective and turbulent water movement among twelve reservoirs, atmospheric exchange among four of those reservoirs, biotic production of organic carbon and calcium carbonate, and rates of oxidation and dissolution, respectively.

Coverage of global diversity, then, can be achieved by studying ten or twenty large world rivers chosen on the basis of drainage area, discharge, and material load. Two or three of these major rivers occur in each of five continents (excluding Australia), spanning wide ranges in vegetation, soils, types and intensities of land use, terrain, and rivers.

Watersheds or river basins function as macro-units, which concisely inte-

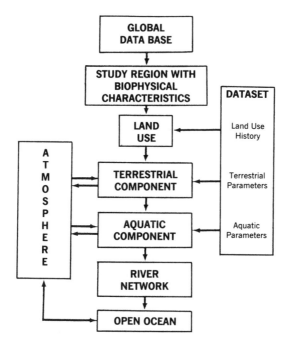

FIGURE V–2   Model macro-structure for global biogeochemical cycling study.

grate the terrestrial, aquatic, and ocean systems. To achieve further geographic resolution within basins and to organize disparate sources of the
necessary data used in a global cycling model, a standardized unit of Earth
surface was selected. Uniform grid cells of longitude and latitude tran-
scend the varying landscape boundaries of soils, vegetation, hydrology,
and the geopolitical boundaries of land use data (county, state, region,
country)—none of which coincide for the same area of land. These grid
cells provide the flexibility and uniformity necessary to integrate this array
of landscape and land use data (Figure V–3). Selection of grid cell size is
primarily based on the global data sets chosen. The units used in current
global data archives of ecological and biophysical characteristics (world
vegetation, world soils, world land use, world evaporation, precipitation,
temperature, and runoff) are commonly 1 × 1 or 0.5 × 0.5 degree grid
cells (Matthews, 1983; Wilson and Henderson-Sellers, 1985; Olson and
Watts, 1982; Baumgartner and Reichle, 1975; Wilmott et al., 1985). We
have chosen the 0.5 × 0.5 degree resolution for our global biogeochemical cycling models.

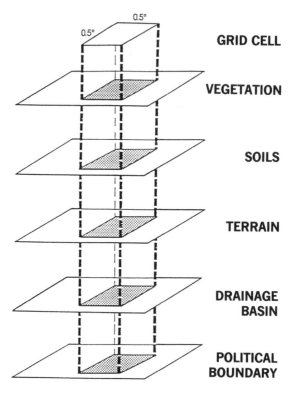

FIGURE V–3   Configuration of data layers within a 0.5 degree latitude by 0.5 degree longitude grid cell.

Each 0.5 × 0.5 degree landscape unit is defined either as a terrestrial cell by its vegetation and soil type or as an aquatic cell representing major water bodies that cover 75 percent of the grid cell. Each terrestrial cell additionally contains an aquatic component, representing stream orders one through six, which drain terrestrial land area within the cell. Each cell is geo-referenced and a topology of grid cell clusters can be constructed in order to simulate material transport. Each cell resides within a watershed or subdrainage basin, and each subbasin is contained within a major river basin. This hierarchical structure provides a unified approach, a mechanism for transfer of information between terrestrial and aquatic modules, and a directional flow of material towards the ocean systems.

To facilitate organizing and integrating the many large data sets with global coverage, we are developing a Geo-referenced Global Data and

Analysis System (G^{2}DAS). This computerized system can present an image that is geographically oriented to a known area on the surface of the Earth and can realign other data sets to that same area, even if they have been obtained at different times and scales or from different sources. The most common sources are maps, remotely sensed data, documentary statistics, and in situ studies.

This information system also allows utilization of data sets with widely different formats. The two most basic are the tessellated and vector formats (Peuquet, 1984). The grid, or regular square mesh, is the most versatile and useful tessellation. Raster formats, characteristic of remotely sensed data, are also widely used. (The differences between raster and grid formats are subtle; for further discussion, see Peuquet, 1984.) The second most frequently encountered data format is the vector, or polygon, format. Encoding data using a vector format maximizes the representation of the original data source in the computer or model. Finally, point data represents another widely used data format, and both regular and irregular point data can be used to develop contoured vector data sets.

Data management is facilitated by dividing a particular data set into its two basic components, the image set and the descriptor set. The image set provides the location information for each geographic unit (i.e., administrative district, stream segment, meteorological station) and stores this in a collection of related data files comprising the data topology. The descriptor set provides the thematic attribute information for each geographic unit of the image data set (i.e., district name, stream order, mean annual precipitation).

Aside from the capacity to couple global data sets to numerical models, the G^{2}DAS itself provides a means for data analysis. One of the most useful applications of this function is the creation of new data sets from combinations of others. The simplest example is change detection analysis, where one data set is subtracted from another, producing a map of areas undergoing change between the two points in time. More complicated analysis can similarly be performed, such as a the computation of a map of soil carbon and nitrogen based on irregularly spaced, in situ studies and data on soil type, elevation, and climate. Data set overlay, distance measurements, cluster analyses, neighbor analyses, network analyses, and spatial statistic are other analytical functions of the G^{2}DAS.

The G^{2}DAS facilitates selection of portions of any data set based on the information contained in the locational (image) or thematic attribute (descriptor) classifications. Through G^{2}DAS, a study region, such as a major world river basin, is selected from its global archive. A data file is formed containing all the 0.5 × 0.5 degree cells comprising the basin. This file contains, for each cell, values for each biophysical parameter later used by the model (e.g., soil, slope, vegetation, subbasin, precipitation, runoff, river size, state, county, region) (Figures V–2, V–3). The data file is combined with data of land use changes for each land use category considered through time for each geographical unit. Following a history of land use changes, the internal dynamics of terrestrial nutrients response and aquatic processing can then be estimated within each grid cell.

A modeling scheme that tracks nitrogen pathways from an atmospheric through a terrestrial and aquatic system to the coastal and open ocean is shown in Figure V–4. A grid cell maintains an inventory of both terrestrial and aquatic N. Inputs into the terrestrial component come from fertilization and/or precipitation. Exports are in the form of nitrification, denitrification, and volatization products. A fraction also leaves in runoff, either through leaching and/or erosion, which serves as input to the aquatic component. Input N from upstream cells is added to the aquatic component and the amount of nutrients lost from the grid cell is determined by a retention coefficient: a function of stream order, impoundment, loading, and biology. This amount is then passed to the next downstream grid cell.

A pictorial example of the history of four adjacent grid cells along a reach of a river is shown in Figure V–5. The relative change in the size of the boxes and arrows indicates the amount of nutrient being tracked. For example, converting a forest to nineteenth-century agriculture (using little or no fertilizer) reduces the quantity of resident terrestrial nutrients. The rivers further act as nutrient processors to retard the movements of nutrients out of the grid cell; material flux out of a grid cell is especially reduced by retention in impoundments. The flux from grid cells containing cities and agriculture with fertilizer is increased. The grid cells within watersheds process both terrestrial and aquatic nutrients, and then connect to major water courses. A range of grid cell responses to land use changes is displayed in Figure V–5. The model easily handles both undisturbed ecosystems and those disturbed by humans.

For initial testing and validation of a biogeochemical model, North America—specifically, the greater Mississippi River drainage basin—was an ideal test case. The Mississippi ranks third globally in drainage area, fifth in discharge, seventh in sediments, and first in nutrient concentration (Meybeck, 1982; Milliman and Meade, 1983). It spans several vegetative communities and major soil groups, and has a wide range of land use types and intensities. Forty percent of the runoff of the continental United States is carried by the Mississippi River (Leopold, 1962). It drains six regional basins (Ohio, Arkansas-White-Red, Missouri, Upper Mississippi,

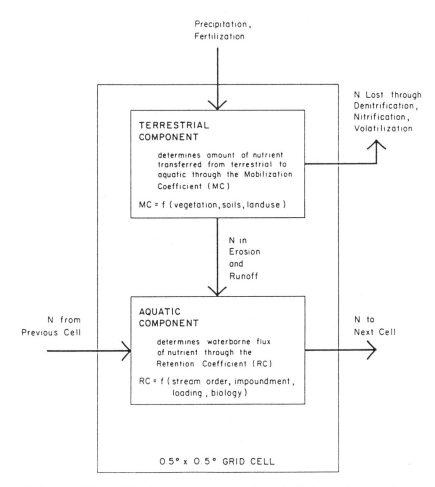

FIGURE V–4   Terrestrial and aquatic components of a grid cell, showing inputs and exports of material.

Lower Mississippi, Texas Gulf) and includes 35 subregional basins as mapped by USGS and U.S. Water Resources Council (1978). The six regional basins touch and/or include 30 states and cover nearly two-thirds of conterminous U.S. land area. Estuarine considerations were put aside initially because the aquatic modeling is limited to fresh water alone (with a similar direct transfer to the 12-box ocean model).

The test case tracked total nitrogen in two contrasting scenarios: an

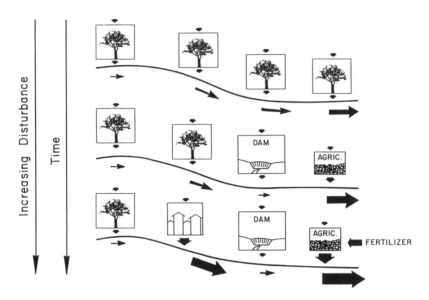

FIGURE V–5 Graphic timeseries of a grid cell and land use change with its effect on nutrient mobilization, retention and transport.

undisturbed, presettlement land cover pattern using the Holdridge Life Zone Classification of potential vegetation (Emanuel et al., 1985) and a contemporary (1975) land use pattern, using state-level statistics on 15 categories of land use. The terrestrial component predicted the average annual total nitrogen lost from each 0.5×0.5 degree grid cell within the Mississippi basin to recipient aquatic systems and to the atmosphere (Gildea et al., 1986; Gildea et al., 1987). The aquatic component (Vorosmarty et al., 1986) collected the nutrient loading, processed nitrogen, and delivered the residual to the coastal ocean. We first present details of the terrestrial component in the Mississippi model and then describe continuing model development strategy for terrestrial nutrient cycling. This is followed by a section on the aquatic component and parallel aquatic nutrient model development.

TERRESTRIAL COMPONENT

An ecosystem's nutrient behavior was characterized by the size of elemental pools and fluxes. Restricting initial modeling efforts to a contrast of contemporary with predisturbance settings not only reduced the number of variables to be considered but also greatly simplified a terrestrial

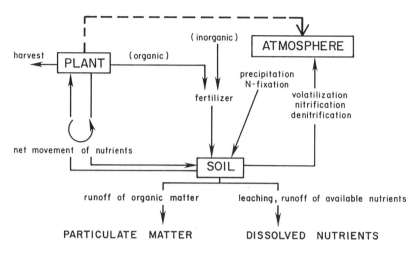

FIGURE V–6   Terrestrial nitrogen budget model (after Frissell, 1977).

nutrient model structure. The terrestrial component initially sidestepped an important limiting condition on a possible model structure: that of changing pool sizes (e.g., declining soil organic matter) due to effects of human management (Jenny, 1941). With disturbed systems at near steady state conditions, characterization of nutrient cycles is reduced to ecosystem net nutrient loss for various types of land use. Furthermore, for the terrestrial component, key forcing functions, such as fertilizer rate and size of yield or forest harvest, showed only spatial and ecosystem-specific differences in a contemporary scenario.

The terrestrial component was represented by a simple nutrient budget (Figure V–6). Modified from Frissel et al. (1977), the model is a donor-controlled, linear, first-order box model and establishes turnover times for a nutrient in the model compartments. Input terms included fertilizer rate by types of land use and by state, N-fixation by ecosystem/land use, and dry and wet deposition by ecosystem/land use. Output terms included harvest/yield by land use, volatilization, and nitrification and denitrification by ecosystem/land use. A leakage term was then produced by difference. The internal fluxes of net plant uptake and plant remains returned to soil (litter), while documented in the Frissel data, were not explicitly treated in the terrestrial submodel. Similarly, livestock droppings, although significant in localized areas, were initially excluded. Any net mineralization or immobilization of soil organic matter that contributed to the coarse N budget became a single output term and was added to the other terms.

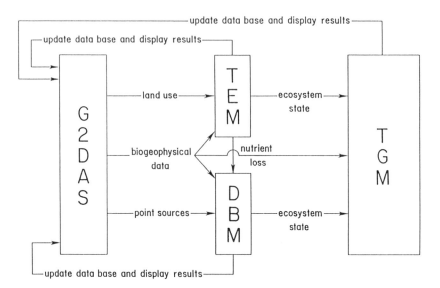

FIGURE V–7   Relationship of grid cell-based models to G²DAS in global biogeochemistry studies.

This model structure, because of the steady state assumptions in pool sizes, could only be used where harvest yield and soil organic matter content have been somewhat constant over the last five to ten years. The model did serve to identify key areas that needed more explicit treatment and development, and to better describe and quantify element movement through large systems and complex landscapes. However, in order to simulate the effects on terrestrial biogeochemical cycles of changing land uses, both historically and for contemporary analyses in developing nations, a more detailed ecological model structure is necessary. Specifically, the internal dynamics of organic matter mineralization and release of inorganic nitrogen, the impact of disturbance on C and N fixation and mineralization, and the seasonality of the significant ecological pathways in element cycling all needed more explicit treatment.

To achieve this, the next series of model development incorporated process-level detail in a three-part approach to element cycling. Three linked, large-system models replaced the original Mississippi model with its simple terrestrial and aquatic nutrient budget components (Figure V–7). The first model, the Terrestrial Ecosystem Model (TEM), describes the ecosystem state of a particular land use or land cover. The Trace Gas Model (TGM) determines the spatial and seasonal patterns of the fluxes of CO_2, CH_4,

CO, and N_2O between terrestrial biomes and the atmosphere, and between aquatic systems and the atmosphere. The Drainage Basin Model (DBM) determines baseline fluxes and recent changes in fluxes of C, N, and P loads from terrestrial biomes in rivers and estuarine systems.

The rates of element movement, relative size of element pools, and the biochemical composition of the pools are indexes that partially characterize the state of an ecosystem. The description of the ecosystem state used in TEM includes net primary production (NPP), heterotrophic respiration, and the stocks of available carbon and nitrogen. In addition, various biogeophysical variables, such as vegetation type, soil type and texture, soil moisture, and temperature changes in biotic activity that occur because of human perturbations and that affect both the exchange of trace gases with the atmosphere and the fluxes of key compounds to the aquatic system, become the linkage among the three models and provide a detailed description of element cycling.

The primary function of TEM is to describe ecosystem state in terms of NPP, respiration, and nutrient availability. Basic geophysical site characteristics are passed from G^2DAS to TEM (Figure V–7). The site indices are in turn used by TGM in the prediction of trace gas fluxes from terrestrial systems and by DBM to predict element movement from terrestrial systems to rivers. The ecosystem state is described by the cycling rates of C and N and, therefore, the potential loss rate of C and N from each system. The cycling rates of C and N in turn become driving variables for the trace gas model (TGM), and the loss rates of C and N become driving variables for the drainage basin model (DBM). TEM includes sufficient process-level dynamics to reproduce NPP from basic abiotic driving variables (light, moisture, temperature), yet is sufficiently aggregated to operate on a global level. That is, it has a reasonable number of parameters and variables, yet is able to tolerate global ecosystem level coarseness.

TEM (Figure V–8) operates at the ecosystem level, aggregating species, ages, and structural parts. CO_2 and available soil nutrients are taken up by the aggregate vegetative biomass at rates controlled by temperature, moisture, and light. CO_2 is respired by the biomass, and similarly controlled by abiotic factors. The biomass releases carbon and nutrients in the form of litter of a specific quality (lignin:N), which decays at specific rates (Aber and Melillo, 1980; Melillo and Gosz, 1983). CO_2 produced by decomposition is influenced by temperature and moisture, with the decay rate and litter quality determining mineralization or immobilization rates of litter. Loss of carbon and nutrients from undisturbed systems is determined by the relationship of available C and nutrients, and wetting events (flooding, leaching). Loss from disturbed systems is determined by the inclusion of

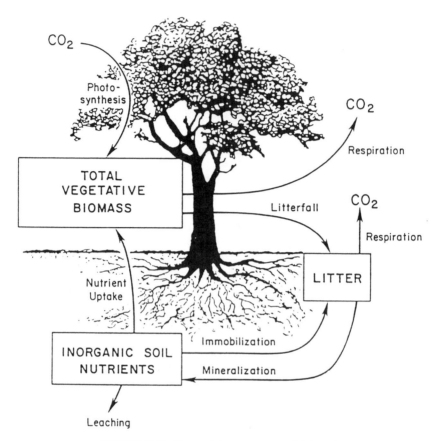

FIGURE V–8   Terrestrial ecosystems model (TEM).

management information, external nutrient inputs (fertilizer), changes in cycling rates due to land use, and wetting events. The model uses functional groups of ecosystems with similar nutrient responses and can follow the development of land use histories within a grid cell, illustrating the shifts from closed, steady state systems through transient systems to open, steady state systems with continuous intensive land use. TEM operates on each terrestrial grid cell for a given month, yielding monthly global estimates of both CO_2 consumption and production, as well as levels of available soil nutrients. Similarly, TGM provides estimates of CO, CH_4, and N_2O fluxes between terrestrial and aquatic biomes and the atmosphere. It has the same temporal and spatial resolution as TEM—monthly to seasonal time steps over the last several decades for each terrestrial 0.5×0.5 degree grid cell globally.

The TGM structure is centered around functional relationships between fluxes of the three trace gases (CO, CH_4, N_2O) and site conditions. From TEM, site conditions are summarized into a site quality index, which is then used to predict trace gas flux rate from a multidimensional response surface. The response surface is developed using relationships from literature on trace gas production under a wide range of environmental conditions for functional groups of ecosystems. The surface is developed for both terrestrial and aquatic functional groups and relates environmental variables identified as most signifcant in influencing gas production rates. The concept of conversion efficiency is used in the biochemical transformations that occur in trace gas production. For example, the conversion efficiency of nitrite to nitrous oxide is a function of the integrated site index that is generated from G^2DAS, TEM, and DEM (Figure V–7). G^2DAS will provide information on geophysical parameters such as soil type, temperature, and moisture conditions, while TEM and DBM will provide information on the biological state of the system, such as the size of the available carbon and nutrient pools. The integrated site index is used to modify a maximum N_2O loss rate reported in the literature.

These models (TEM, TGM) represent new levels of specificity in describing ecological consequences of changing global land use. By incorporating ecosystem processes that can predict changes in the pattern of net primary production, respiration, and nutrient availability and, therefore, resultant changes in the distribution of CO_2 and other trace gas exchanges, significant changes in terrestrial ecosystem element cycling can more clearly be understood. We have moved through a similar evolution in treating element cycling in recipient aquatic ecosystems with increasing levels of detail and specificity.

AQUATIC COMPONENT

Any model of recipient aquatic ecosystems must support three interacting subcomponents of waterborne nutrient cycling: hydrology, mass balance, and active nutrient exchange. Methods for computing volumes of water that can contribute to river flow are highly advanced at the local scale (Haan et al., 1982; Singh, 1982). Fewer attempts have been made at the global level (Baumgartner and Reichle, 1975; Korzoun et al., 1977), and we know of but one model that routes entire drainages at such a scale (Gianessi et al., 1981). Because we consider the movement of biotically active materials within large drainages, both the effect of riparian zone nutrient kinetics and prolonged residence within aquatic ecosystems must be determined. The attendant effects of metabolism in water bodies of varying size and characters, such as small streams, floodplain rivers, and artificial reservoirs, are important determinants of the ultimate delivery of nutrients to the coastal ocean. In reality, it is the combined effect of numerous and complex mechanisms that constitutes nutrient processing within riparian and aquatic ecosystems. These include plant growth, diffusion into and out of sediments, gaseous losses, chemical transformation, sedimentation, and resuspension.

For the purposes of constructing a regional model, substantial simplification was required. Our strategy used a simple retention/source coefficient that represents the net removal or release of material due to biotic, chemical, or physical agents. The coefficient was expressed as a proportion of nutrient loading. Although simple in concept, the data required to parameterize large-scale models operating over long time steps are extremely limited. A view that emerges (Correll, 1986) indicates that riparian zones, reservoirs and other impoundments, and small rivers retain or lose a significant proportion of their incident nutrient loads. Large rivers are more passive.

The strategy for a regional aquatic ecosystem model and results for the Mississippi test case are described in Vorosmarty et al. (1986). The model operated at an annual time step using a spatial grid of 0.5 × 0.5 degrees. Because of the coarse time resolution, transport based on water flows was inappropriate. Instead, nutrient mass was tracked directly from cell to cell. The basic accounting unit was the assessment subregion as defined by the United States Water Resources Council (1978). The Mississippi drainage comprised 35 such subregional basins, each representing a collection of contiguous, geo-referenced, 0.5 × 0.5 degree grid cells. Information on direction of flow and location of reservoirs was digitized from State Hydrologic Unit Maps prepared by the USGS (1976). Nutrients generated from terrestrial ecosystems within each basin were collected, sequentially passed through small streams and rivers, and processed before entering larger water courses. Processing of nutrients was based on stream size, determined from well-established geomorphologic relationships linking drainage area to river geometry (Jarvis and Woldenberg, 1984; Leopold et al., 1964) and field studies of nutrient cycling in rivers and impoundments of differing size (see Vorosmarty et al., 1986). Simple mass balance was used to predict the delivery of material to a major river draining the basin. The resulting flux was passed to the adjacent downriver basin. By so accounting for all nutrient mass within the Mississippi catchment, the annual delivery of material to the coastal ocean was estimated.

Employing such single basin studies to attain full global coverage may be a long-term goal of global ecology studies. However, a careful sampling of study regions can lend more immediate insight into the impact that humans have on biogeochemistry at the global level. Table V–2 shows some large and important drainage basins of the world. The 11 basins listed represent contrasts between nonindustrial and industrial agriculture, low and high population density, arid and humid climate, and tropical and arctic regions. Collectively, they drain a significant portion, about 30 percent, of the continental surface area that delivers runoff to the world's oceans. They also represent 30 percent of all water flow and sediment discharge to the oceans. These major rivers are enriched with respect to dissolved N and P flux, and constitute 40 percent or more of the current global total for each nutrient. When compared to the global anthropogenic load, this subset of rivers is further enriched—at least 60 percent and 50 percent of the added N and P loads, respectively. An analysis of these drainage basins using our model, therefore, would cover a substantial fraction of the Earth's surface subject to active human disturbance of nutrient cycles.

Basins such as these will be analyzed through an enhanced version of the Mississippi River model described earlier using the Drainage Basin Model I (DBM I) shown in Figure V–9. DBM I predicts runoff volumes from meteorologic data and automates water and nutrient routing using subannual (i.e., seasonal or monthly) time steps. Runoff is determined using the water balance techniques developed by Thornthwaite and Mather (1957) and Dunne and Leopold (1978). Runoff represents the balance among rainfall, actual evapotranspiration, and changes in soil moisture. To predict excess water that can leave a grid cell, it is necessary to define the average moisture-holding capacity of a soil. This is inferred from soil texture using our FAO soils data base (Gildea and Moore, 1986) and rooting depth from a global vegetation map, such as that provided by Matthews (1983). Next, meteorologic data in the form of monthly precipitation and potential evapotranspiration is specified. Actual evapotranspiration, soil moisture stocks, and runoff are then predicted. River networks are established using a global digital terrain data base (Department of Navy). Flow pathways from any particular grid cell are predicted by examining the elevation of adjacent cells and determining the direction of maximum gradient. Cumulative flows can be constructed from this routing. Flow rates are modified by reservoirs, either through enhanced evaporative loss or through human control.

Terrestrial nutrient loading is calculated by TEM and from digitized locations of point sources. Retention/source coefficients are assigned,

TABLE V-2 CHARACTERISTICS OF 11 MAJOR RIVERS OF THE WORLD

River Basin	Area(a) km² × 10	Discharge(a) m³/sec	Sediment(b) Load 10⁶ l/yr	Dissolved(c) Inorganic N Load 10⁹ g/yr	Phosphate(c) P Load 10⁹ g/yr
Amazon	6,300	175,000	900	420	65
Zaire	4,000	39,200	43	120	30
Mississippi	3,250	18,400	210	640	120(g)
Nile	3,000	2,800	0	NA	NA
Parana	2,800	18,000	92	280	40
Tenisel	2,600	17,200	13	NA	NA
Yangtze	1,950	22,000	478	190	NA
Orinoco	950	30,000	210	90	5
Ganges/Brahmaputra	1,550	30,900	1,670	340	50
Makong	800	18,300	160	140	NA
Rhine	150(d)	2,000(d)	NA	280	15
Above Total	27.4 × 10³	373.8 × 10³	>3.8 × 10³	>2.5 × 10³	>325
Contemporary Continental Total	100 × 10³(e)	1,175 × 10³(f)	13.5 × 10³	6.5 × 10³(h)	800

134

Share of the World Total by Rivers Listed Above	27%	32%	>28%	>38%	>41%
Natural Transport (Global)				4.5×10^3	400
Anthropogenic Nutrient Source (Global)				2×10^3(h)	400
Fraction of Source Attributable to Rivers Listed Above				>50%(i)	>60%(i)

(a) Meybeck (1967).

(b) Milliman and Meade (1983). The load reported is that delivered to world oceans.

(c) Data for inorganic N are for sum of nitrate-N, nitrite-N, and ammonium-N, as available. Loads calculated using discharge from Meybeck (1976) and concentration from Meybeck (1982).

(d) van Bennekom and Salomons (1981).

(e) Meybeck (1982) and Alekin and Brazhnikova (1960). Area refers to that which delivers discharge to the world's oceans.

(f) Average of Baumgartner and Reichel (1975) and Alekin and Brazhnikova (1960). Discharge is that flowing into world oceans.

(g) Based on the ratio of PO_4-P to total dissolved P of 0.4 as given by Meybeck (1982).

(h) Based on the ratio of dissolved inorganic N to total dissolved N of 0.30 (Maybeck, 1982).

(i) We assumed that 30% of the natural load is attributable to the 11 major rivers listed above in a predisturbance condition. Any additional load was assigned to human disturbance in these basins, and the collective load was then compared to the global total.

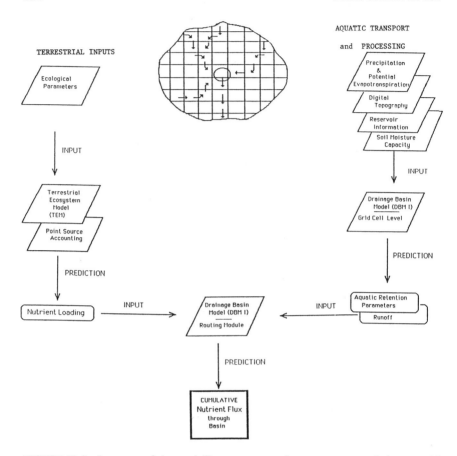

FIGURE V–9 Summary of the modelling strategy used to generate cumulative material loads within large drainage basins

using information on stream size (a function of the landscape area drained) and the presence of lakes or reservoirs within the grid cell. Areas of low relief have a correspondingly greater proportion of riparian vegetation and hence intercept more of the upland nutrient export. Nutrients are carried with predicted river flows and a cumulative export for the entire basin emerges. Model predictions are then validated against existing data, such as Meybeck's (1982) excellent global summary.

In theory such a model could be used not only for examining long-term average conditions, but transient hydrologic events as well. Probability functions for precipitation and evapotranspiration could be developed and applied to predict the impact of major flooding or prolonged drought.

Such simulations could describe historical events or explore the consequences of future climate change.

A second Drainage Basin Model (DBM II) is required in order to achieve global coverage of riverine water, carbon, and nutrient fluxes. Ecologists and biogeochemists have successfully predicted the export of nutrients from watersheds, using information on geology, ecosystem type, land use, human energy use, and population (Meybeck, 1982; Dillon and Kirchner, 1975). For example, Meybeck (1982) found that nitrogen and phosphorus export in rivers of various countries increased as a function of both population density and per capita energy consumption.

A Demophoric Index, defined as the ratio of annual per capita energy consumption to minimum energy required to satisfy human needs (Vallentyne, 1978), correlates well with both excess phosphorus and excess nitrogen export from drainage basins. For example, in highly developed countries, such as the United States and Germany, total dissolved excess nitrogen exports are as high as 7 to 10 kg N per person per year, while in less developed regions, such as Java, exports are as low as 0.5 kg N per person per year. These annual estimates would need to be partitioned seasonally, linked to runoff or seasonal patterns of nutrient use. A global compilation of land use and population will supply the demophoric information needed to predict continental export of biotically important nutrients and carbon. In linking DBM II to TGM, CO_2 and trace gas production can be determined from aquatic ecosystems within each drainage basin. Predictions of gas flux are based on nutrient status, carbon loading, degree of anoxia, and temperature.

SUMMARY

A strategy for understanding changes in the global cycling of biologically important elements has been presented. To illustrate this strategy, a detailed model for the greater Mississippi drainage basin was developed. It integrated temporal and spatial variation in land use and ecosystem response that had not previously been attempted. While clearly recognizing the potential problems in some of the necessary assumptions, we nevertheless set forth a framework for determining fluxes at a high degree of spatial resolution over large regions. In previous global models, such as those of terrestrial carbon (Moore et al., 1981; Houghton et al., 1983), spatial resolution or regional specificity was confined to entire continents or other huge areas. Global climatological models, such as the general circulation models, utilize a grid cell approach to describing the Earth's surface (Hansen et al., 1983), as do global hydrological models (Baumgartner and Reichle, 1975). However, none exist that integrate both terrestrial and aquatic ecosystems with high resolution globally.

While many of the internal relationships of the model were developed using the Mississippi test case, the development of detailed process models (TEM, TGM, and DBM I and II) furthers our approach to understanding changes in the metabolic system of the planet. The structure is sufficiently flexible to handle global differences in vegetation and soil, agricultural technology and intensity, crop type, and river morphology. It also establishes the first integrated framework for characterizing nutrient flow dynamics and identifying the impacts of human activity at the global scale. We expect it to function as an incisive mechanism for testing the adequacy of our knowledge of biogeochemical cycles. We trust it will become fertile ground for identifying avenues of future research and investigation.

In conclusion, the strategy presented here has been to design a model that functions globally with a reasonably high degree of partial resolution of types of landscapes. A logical mechanism for integration of Earth's landscapes, the watershed, is used to process and transport nutrients of terrestrial origin and deliver them to ocean systems. The structure utilizes temporal and spatial variation in land use within representative terrestrial ecosystems to mobilize nutrients and to deliver them to freshwater aquatic ecosystems. Anthropogenic impacts are determined by contrasts between pre- and postdisturbance scenarios and by documenting the nutrient "signals" found in the world's river systems. Additionally, the structure creates atmospheric point source terms that can link to other global models of general circulation studies.

The understanding gained from these modeling exercises will be used to examine contemporary and possible future trends. These models function as a link between detailed ecological community process models and models of atmospheric chemistry and circulation. We view these models as a way of scaling up from the small landscape unit to the globe with a high degree of geographic specificity. Such models not only yield insights into the regional aspects of global biogeochemical problems, but also clearly illustrate areas that need further investigation by the scientific community.

REFERENCES

Aber, J. and J. Melillo. 1980. Litter decomposition: measuring relative contributions of organic matter and nitrogen to forest soils. *Canadian Journal of Botany* 58, 4:416–421.

Alekin, O.A. and L.V. Brazhnikova. 1960. A contribution on runoff of dissolved substances on the world's continental surfaces. *Gidrokhimicheskie Materialy* **32**:12–34.

Bacastow, R.B. and C.D. Keeling. 1981. Atmospheric carbon dioxide concentration and the observed airborne fraction. In: *Carbon Cycle Modeling*, B. Bolin, ed. SCOPE 16. John Wiley and Sons, New York. Pp. 103–112.

Banin, A.J., J.G. Lawless, and R.C. Whitten. 1984. Global N_2O cycles, terrestrial emissions, atmospheric accumulation, and biospheric effects. *Advances in Space Research* 4, 12:207–216.

Baumgartner, A. and E. Reichel. 1975. *The World Water Balance*. Elsevier Scientific Publishing Company, Amsterdam, 179 pp.

Bolin, B. 1981a. *Carbon Cycle*, SCOPE 16. John Wiley and Sons, New York.

Bolin, B. 1981b. *Changing Global Biogeochemistry*. Report CM–52. Department of Meteorology, University of Stockholm.

Bolin, B., A. Bjorkstrom, and B. Moore. 1983. The simultaneous use of tracers for ocean circulation studies. *Tellus* 35:206–236.

Correll, D., ed. 1986. *Watershed Research Perspectives*. Smithsonian Institution Press, Washington, DC, p. 421.

Dillon, P.J. and W.B. Kirchner. 1975. The effects of geology and land use on the export of phosphorus from watersheds. *Water Research* 9:135–148.

Emanuel, W.R., H.H. Shugart, and M.P. Stevenson. 1985. Climatic change and the broad-scale distribution of terrestrial ecosystem complexes. *Climatic Change* 7:29–43.

Frissel, M.J., ed. 1977. Cycling of mineral nutrients in agricultural ecosystems. *Agro-Ecosystems* 4:1–354.

Gianessi, L.P., H.M. Peskin, and G.K. Young. 1981. Analysis of National Water Pollution Control Policies. *Water Resources Research* 17, 4:796–801.

Gildea, M.P. and B. Moore. 1986. Development and management of a digital global soils data base. Presented at Canadian Soil Science Society meeting, July 7–10, 1986, Saskatoon, Saskatchewan, Canada.

Gildea, M.P., B. Moore, C.J. Vorosmarty, J.M. Melillo, K. Nadelhoffor, and B.J. Peterson. 1987. Impacts of land use on nitrogen cycling in large systems: The Mississippi River basin. Presented at United States Association of Landscape Ecology, University of Virginia, Charlottesville, Virginia, March 9–11, 1987.

Gildea, M.P., B. Moore, C.J. Vorosmarty, B. Bergqinst, J.M. Melillo, K. Nadelhoffor, and B.J. Peterson. 1986. A global model of nutrient cycling: I. Introduction, model structure, and terrestrial mobilization of nutrients. In: *Watershed Research Perspectives*, Smithsonian Institution Press, Washington, DC, pp. 1–31.

Haan, C.T., H.P. Johnson, and D.L. Brakensiek, eds. 1982. *Hydrologic Modeling of Small Watersheds*. American Society of Agricultural Engineering, St. Joseph, Michigan, p. 533.

Hansen, J., G. Russel, D. Rind, P. Stone, A. Lacis, S. Lebeduff, R. Ruedy, and L. Travis. 1983. Efficient three-dimensional global models for climate studies: Models I and II. *Monthly Weather Review* 3:609–662.

Houghton, R.A., J.E. Hobbie, M.M. Melillo, B. Moore, B.J. Peterson, G.R. Shaver, and G.M. Woodwell. 1983. Changes in the carbon content of terrestrial biota and soils between 1860 and 1980: A net release of CO_2 to the atmosphere. *Ecological Monographs* 3:235–262.

Houghton, R.A., W.H. Schlesinger, S. Brown, and J.F. Richards. 1985. Carbon dioxide exchange between the atmosphere and terrestrial ecosystems. In: *Atmosphere Carbon Dioxide and the Global Carbon Cycle*, J. Trabalka, ed. U.S. Dept. of Energy, DOE/ER–0239, Washington, DC.

Jarvis, R.S. and M.J. Woldenberg, eds. 1984. *River Networks*. Hutchinson Ross Publishing Company. Benchmark Papers in Geology, vol. 80.

Jenny, H. 1941. *Factors of Soil Formation*. McGraw-Hill Book Co., New York.

Johnson, P.L. and W.T. Swank. 1973. Studies of cation budgets in the Southern Appalachians on four experimental watersheds with contrasting vegetation. *Ecology* **54**:70–80.

Khalil, M.A.K. and R.A. Rasmussen. 1984. Carbon monoxide in the Earth's atmosphere: Increasing trend. *Science* **224**:54–56.

Khalil, M.A.K. and R.A. Rasmussen. 1983a. Sources, sinks, and seasonal cycles of atmospheric methane. *Journal of Geophysical Research* **88**:5131–5144.

Khalil, M.A.K. and R.A. Rasmussen. 1983b. Increase and seasonal cycles of nitrous oxides in the Earth's atmosphere. *Tellus* **363**:161–165.

Korzoun, V.I., A.A. Sokolov, M.I. Budyko, K.P. Voskrensensky, G.P. Kalinin, A.A. Konoplyantsev, E.S. Korotkevich, and M.I. Lvovick. 1977. *Atlas of World Water Balance*. UNESCO, Paris.

Likens, G.E., F.H. Bormann, R.S. Pierce, J.S. Eaton, and N.M. Johnson. 1977. *Biogeochemistry of a Forested Ecosystem*. Springer-Verlag, New York, 146 pp.

Leopold, L.B., ed. 1962. Rivers. *American Scientist* **50**:511–537.

Leopold, L.B., M.G. Wolman, and J.P. Miller. 1964. *Fluvial Processes in Geomorphology*. W.H. Freeman and Company, San Francisco, p. 522.

Matthews, E. 1983. Global vegetation and land use: New high resolution data bases for climate studies. *Journal of Climate and Applied Meteorology* **2**:474–487.

Melillo, J.M. and J.R. Gosz. 1983. Interactions of biogeochemical cycles in forest ecosystems. In: *The Major Biogeochemical Cycles and Their Interactions*, B. Bolin and R.B. Cook, eds. SCOPE 21, John Wiley and Sons, New York, pp. 177–222.

Meybeck, M. 1976. Total mineral dissolved transport by world major rivers. *Hydrological Society Bulletin* **21**:265–284.

Meybeck, M. 1982. Carbon, nitrogen, and phosphorus transport by world rivers. *American Journal of Science* **282**:401–450.

Milliman, J.D. and R.H. Meade. 1983. World-wide delivery of river sediment to the oceans. *Journal of Geology* **91**:1–21.

Moore, B. 1985. Fossil fuels: Carbon dioxide in the atmosphere and oceans. In: *Wastes in the Ocean*. Volume 4. *Energy Wastes in the Ocean*. John Wiley and Sons, New York.

Moore, B. and A. Bjorkstrom. In press. Developing parameters for ocean carbon models by the inverse method. *Proceedings of the Sixth ORNL Life Sciences Symposium*. Knoxville, Tennessee.

Moore, B. and B. Bolin. 1987. The oceans, carbon dioxide, and global climate change. *Oceanus* **29**:9–15.

Moore, B., H. Morowitz, and M.N. Dastoor. 1984. Biogeochemical cycles: Interactions in global metabolism. In: *The Interactions of Global Biogeochemical Cycles*, B. Moore and M.N. Dastoor, eds. NASA-JPL Publication 84-21, pp. 17–24.

Moore, B., R.D. Boone, J.E. Hobbie, R.A. Houghton, J.M. Melillo, B.J. Peterson, G.R. Shaver, C.J. Vorosmarty, and G.M. Woodwell. 1981. A simple model for analysis of the role of terrestrial ecosystems in the global carbon budget, pp. 365–385. In: *Modelling the Global Carbon Cycle*. SCOPE 16. B. Bolin, ed. John Wiley and Sons, New York.

Olson, J. and J.A. Watts. 1982. Carbon in live vegetation of major world ecosystems. ORNL-5862. Oak Ridge National Laboratoy, Oak Ridge, Tennessee.

Peterson, B.J. 1981. Perspectives on the importance of the oceanic particulate flux in the global carbon cycle. *Ocean Science and Engineering* **6**:71–108.

Peterson, B.J. and J.M. Melillo. 1985. The potential storage of carbon caused by the eutrophication of the biosphere. *Tellus* **37B**:117–127.

Peuquet, D. 1984. A conceptual framework and comparison of spatial data models. *Cartographica* 21, 4:66–113.

Rotty, R. 1981. Data for global CO_2 production from fossil fuels and cement. In: *Modelling the Global Carbon Cycle*, B. Bolin, ed. SCOPE 16, John Wiley and Sons, New York, pp. 121–126.

Sagan, D., ed. 1985. *The Global Sulfur Cycle*. NASA Technical Memorandum 87570. Washington, DC 20546.

Singh, V.P., ed. 1982. *Rainfall-Runoff Relationship*. Water Resources Publications, Littleton, Colorado, p. 582.

Steudler, P.A. and B.J. Peterson. 1985. Annual cycle of gaseous sulfur emissions from a New England *Spartina alterniflora* marsh. *Atmospheric Environment* 19:1411–1416.

Thornthwaite, C.W. and J.R. Mather. 1957. *Instructions and Tables for Computing Potential Evapotranspiration and the Water Balance*. Drexel Institute of Technology. Publications in Climatology, Vol. 10, No. 3, pp. 185–303.

U.S. Geological Survey. 1976. State Hydrologic Unit Maps. USGS, Reston, Virginia.

U.S. Water Resources Council. 1978. *The Nation's Water Resources 1975–2000*. Volumes 1–4. U.S. Government Printing Office, Washington, D.C.

Vallentyne, J.R. 1978. Today is yesterday's tomorrow. *International Association of Theoretical and Applied Limnology* 20:1–20.

van Bennekom, A.J. and W. Salomons. 1981. Pathways of nutrients and organic matter from land to oceans through rivers. In: *River Inputs to Ocean Systems*, UNIPUB, New York, pp. 33–46.

Vorosmarty, C.J., M.P. Gildea, B. Moore, B.J. Peterson, B. Bergquist, and J.M. Melillo. 1986. A global model of nutrient cycling: II. Aquatic processing, retention, and distribution of nutrients in large drainage basins. In: *Watershed Research Perspectives*, Smithsonian Institution Press, Washington, DC, pp. 32–56.

Wilmott, C.J., C.M. Rowe, and Y. Mintz. 1985. Climatology of the terrestrial seasonal water cycle. *Journal of Climatology* 5:589–606.

Wilson, M.W. and A. Henderson-Sellers. 1985. A global archive of land cover and soils data for use in general circulation climate models. *Journal of Climatology* 5, 2:119–143.

VI GLOBAL ECOLOGICAL RESEARCH AND PUBLIC RESPONSE

M.B. RAMBLER AND L. MARGULIS

Environmental concern of human impacts and policy formulation first took hold in the United States in the late 1800s. Environmental activists became aware of the relationship of people to natural resources as vast regions of land in the western United States were converted to agricultural use. The dangers of resource exhaustion and misuse were transmitted to the White House, culminating in the establishment of national parks and forests in the early 1900s. As environmental concerns grew stronger in the 1930s, demands were made for the government to take responsibility for deteriorating resources. The government responded to public pressure by instituting the Soil Conservation Service, various watershed programs in the Forest Service, the Tennessee Valley Authority, and the Bureau of Reclamation. These initiatives continued through the 1950s and 1960s with little attention devoted to the consequences of further urban and agricultural development encroaching on national resources.

A shift in emphasis from conservation to the maintenance of environmental quality occurred in the late 1960s and the 1970s. Numerous regulations and policies such as clean air and water acts were imposed on the public by the Environmental Protection Agency.

Perhaps it was the image of the Earth (Figure VI–1) that led to what seems to have been a quantum leap by scientists and environmental action groups. The United States Environmental Program, the 1972 Stockholm

FIGURE VI–1   Earthrise from the Moon.

Conference, the ICSU (Malone and Roerderer, 1985), and many other science symposia and public forums have begun to think of the Earth as a whole, a dynamically functioning collection of ecosystems and biomes connected through atmosphere, oceans, and sediments. Attention is now being focused on the fundamentals of the system that supports life on Earth: mineral cycling, ocean and terrestrial productivity, perturbation effects, and biospheric responses. This new awareness fostered a new approach to the science of the biosphere, a holistic approach to Earth. It has led, for example, to the formation of GERO (Global Environmental Research Organization), an international effort whose five-item agenda we append here (see Appendix II, pp. 171-184). As the goals of GERO make clear, satellite remote sensing and data handling advancements have placed this objective within our grasp.

The installation, within NASA's Life Science and Planetary Science divisions, of the global habitability program has been instrumental in providing leadership for this incipient global ecological research program.

The recognition that life has exerted a geological and atmospheric force on the Earth's surface is a relatively new concept. There has been such an intimate level of interaction of life with its planetary environment that books can no longer be written about the history of the climate, atmosphere, or sediments without a consideration of the history of life (Walker, 1986; Schneider and Londer, 1984). Atmospheric chemists, microbiologists, geologists, and even mathematicians are becoming global ecologists.

Conventionally, ecologists have dealt with isolated segments of the biosphere, called ecosystems—desert habitats, temperate grasslands, tropical forests, and so forth. Only in the last few years has there been a recognition of a dynamic Earth where the biota is inextricably linked to atmospheric, oceanic, and terrestrial processes, where ecosystems are connected to globally by the atmosphere, oceans, and sediments.

The biota affects both short-term events that occur on annual or decadal time scales as well as long-term events, occurring over millenia. Scientists investigating chemical/physical aspects of the planet have recognized that biological processes drive some of the most important geochemical and atmospheric cycles. But the details of even the major biogeochemical cycles are still virtually unknown. Biologists have recognized the need to integrate information concerning biological processes (e.g., growth, metabolism, excretion, etc.) with that concerning atmospheric, oceanic, and terrestrial processes in order to understand the fundamentals of the system that supports life on Earth. Such integration requires the approach taken here of ecology from a global perspective.

Until recently we have not had the appropriate tools to deal with problems on a global scale. However, new capabilities have developed by virtue of high-resolution satellites, highly sensitive analytical instrumentation, and sophisticated computers.

A program of global ecological research, whether led by GERO, NASA, or university researchers, is envisioned to include efforts in modeling, field, and laboratory investigations. Modeling would be directed along two lines: one dealing with interactive aspects of biology and the physical Earth in a more or less static manner, the other dealing with dovetailing the models of individual biogeochemical cycles. Eventually the models would converge, integrating both the interactive aspect and the dynamics. Modeling efforts would also designate areas of uncertainty for future research direction and emphasis. Such future directions might include measurement of the Earth by remote sensing, aircraft, and satellite field measurements to establish ground correlations to remote sensing data and laboratory investigation on rates and specific aspects of biogeochemical pathways.

Because of recently recognized practical needs, which we and other nations of the world will encounter in the coming decades, it is important that we pursue this global view of the Earth's biota to its fullest extent. An understanding of biogeochemical processes has been initiated in other agencies and in other countries. However, NASA has a vital role in providing to the public both a planetary perspective, allowing us to deal intelligently with oceans, polar ice caps, atmospheric and climate changes, volcanic events, and so forth, as well as Earth observation generated by satellites such as Landsat, UARS, ERBE, and so on. An essential component in this initiative is the involvement of biologists in ecological aspects of biogeochemical cycling and for the development of a balanced interdisciplinary program that has a planetary context.

The Gaia hypothesis has generated great interest and excited response in the reading public (Lovelock, 1979; Myers, 1984; Snyder, 1985; White, 1987). It has even inspired some television programs (e.g., Goddess of the Earth, BBC and NOVA versions) and lecture series (e.g., at the Cathedral of St. John the Divine, New York City). A Gaia publishing house has been formed (i.e., Gaia Books, London), and plans to form a Gaia foundation are under way. If we can translate public interest in the Gaia hypothesis into support that is not just financial for scientific research on the global scale, perhaps the day when human beings treat the Earth as a shared, living planet may draw somewhat closer.

SUMMARY

Unlike the other planets of our solar system, the Earth has not traditionally been studied as a single entity. The remedy is in sight as a new field of global science emerges. We planned this book of integrated readings to inhibit the march of academic apartheid. So often the public deals with whole Earth issues piecemeal and on an *ad hoc* basis. Stimulated by popular culture, the responses to global problems tend to be in a crisis mode. However, we have been rejuvenated and directly stimulated by the magnificent image of the living Earth from space. We insist on rigorous science as we witness a shift in point of view within the scientific community. No geologist, biologist, space scientist, or climatologist can any longer work in isolation. All of us must continue to recognize the inextricable linkage among solar, atmospheric, oceanic, and surface processes modulated by life. This book provides an introduction to global concepts through readings, a key bibliography, and research activities in global ecology. As guide both to teaching and research in this newly emerging field, we provide a scientific context for the inevitable planetary perspective of the twenty-first century.

REFERENCES

Lovelock, J. 1979. *Gaia: A New Look at Life on Earth*. Oxford University Press, Oxford.
Malone, T.F. and J.G. Roederer, eds. 1985. *Global Change*. The proceedings of a symposium

sponsored by the International Council of Scientific Unions (ICSU) during its 20th General Assembly in Ottawa, Canada, on 25 September 1984. Cambridge University Press, Cambridge.

Margulis, L. and D. Sagan. 1986. *Microcosmos: Four Billion Years of Evolution from Our Bacterial Ancestors.* Summit Books, New York.

Myers, N. 1984. *Gaia: An Atlas of Planet Management.* Anchor Press/Doubleday and Company, Garden City, New York.

Rambler, M.B., ed. 1983. *Global Biology Research Program. Program Plan.* NASA Technical Memorandum 85629. NASA Scientific and Technical Information Branch, Washington, DC.

Schneider, S.H. and R. Londer. 1984. *The Coevolution of Climate and Life.* Sierra Books, San Francisco.

Snyder, T.P., ed. 1985. *The Biosphere Catalogue.* Synergetic Press, London and Fort Worth.

Vernadsky, V.I. 1986. *The Biosphere* (an abridged version based on the French edition of 1929). Synergetic Press, London and Fort Worth.

Walker, J.C.G. 1986. *Earth History: The Several Ages of the Earth.* Jones and Bartlett Publishers, Boston and Portola Valley.

White, F. 1987. *The Overview Effect.* Houghton Mifflin Co., Boston.

I GLOSSARY OF TERMS AND ACRONYMS

ADCLS   Automated Data Collection and Location Systems; Sensor subsystem on EOS (q.v.) platform

AgRISTARS   Agricultural and Resource Inventory Surveys Through Aerospace Remote Sensing

airshed   A relatively homogeneous air mass with a predictable circulation pattern

albedo   That fraction of the total light incident on a reflecting surface, especially a celestial body, which is reflected back in all directions

altruism   Self-sacrifice for the benefit of others; any form of behavior that increases the fitness of the recipient while reducing the fitness of the altruistic individual

animal   Member of the kingdom Animalia. Eukaryotes that develop from diploid blastular embryos derived from anisogamous (egg/sperm) fertilization

APACM   Atmospheric Physical And Chemical Monitor; sensor subsystem on EOS (q.v.) platform

arroyo   Usually in the arid and semiarid regions of southwest U.S. (a) The small, deep, flat-floored channel or gully of an ephemeral stream or of an intermittent stream, usually with vertical or steeply cut banks of unconsolidated material at least 60 cm high. Arroyos are usually dry, but may be transformed into a temporary water course after heavy rainfall. (b) The small, discontinuous, intermittent stream, brook, or rivulet that occupies such a channel

asexual reproduction   The process that augments the number of cells or organisms deriving from a single parental cell or organism

atmosphere   The mixture of gases that surrounds a planet (e.g., on the Earth the atmosphere is composed of approximately 79 percent nitrogen (N_2) and 21 percent oxygen (O_2), with smaller quantities of argon, carbon dioxide, water vapor, hydrogen, methane, sulfur oxides, and other gases)

atoll   A ring-shaped coral reef appearing as a low, roughly circular (sometimes elliptical or horseshoe-shaped) coral island or a ring of closely spaced coral islets encircling or nearly encircling a shallow lagoon in which there is no preexisting land or islands of noncoral origin and surrounded by deep water of the open sea; it may vary in diameter from 1 km to more than 130 km and is especially common in the western and central Pacific Ocean

autopoiesis   The organizing principle of life by which an entity's boundary structure (e.g., cell membrane) and processes of metabolism and energy exchange are determined by the entity's internal organization and its dynamic interchange with its immediate surrounding

autopoietic entity   Bounded structure displaying principle of autopoiesis (q.v.)

autotroph   Nutritional mode of organisms that can form their own macromolecules from inorganic compounds (such as carbon dioxide) and obtain energy from other inorganic compounds or light; an organism requiring no preformed organic compounds in the diet as sources of energy or carbon (see Table AI–1, p. 151)

AVHRR   Advanced Very High Resolution Radiometer

bacteria   Cellular organisms that lack membrane-bounded nuclei, (i.e., prokaryotes (q.v.)). About 10,000 "species" are known, including multicellular forms such as cyanobacteria (blue-green algae) and actinobacteria (actinomycetes) (see Starr et al., reference on p. 49).

binary fission   An amitotic, asexual division process by which a parent prokaryote cell splits transversely into daughter cells of approximately equal size

biomass   The quantity of living matter in a particular volume, expressed as total weight (mass) of the living organisms per unit area or per volume of the environment

biome   A major terrestrial or marine ecological region (e.g., tropical rainforest biome, tropical marine biome, desert biome, tundra biome, coniferous forest biome, etc.). A set of continuous ecosystems (q.v.)

biomineralization   Formation of minerals influenced by the metabolic processes of organisms (see Table AI-2, pp. 152-156.)

biosphere   The area at the surface of the Earth occupied or favorable for

TABLE AI-1. MODES OF NUTRITION: A list of the sources of energy, electrons and carbon for metabolism; name of each mode with examples of growth of organisms to which names apply. Names constructed by addition of suffix "-troph," e.g., photolithautotroph (plants).

ENERGY light or chemical compounds	ELECTRONS (or hydrogen donors)	CARBON donors	ORGANISMS and their hydrogen (proton/ electron) donors
PHOTO- (light)	LITHO- (inorganic compounds and C_1)	AUTO- (CO_2)	PROKARYOTES: Chlorobiaceae, H_2S, S Chromatiaceae, H_2S, S Rhodospirillaceae, H_2 cyanobacteria, H_2O chloroxybacteria, H_2O PROTOCTISTA (algae), H_2O PLANTS, H_2O
		HETERO- (CH_2O)	NONE
	ORGANO- (organic compounds)	AUTO-	NONE
		HETERO-	PROKARYOTES: Chromatiaceae, org. comp. Chloroflexaceae, org. comp. Heliobacteriaceae, (?) *Rhodomicrobium*, C_2, C_3
CHEMO- (chemical compounds)	LITHO-	AUTO-	PROKARYOTES: methanogens, H_2 hydrogen oxidizers, H_2 methylotrophs, CH_4, $CHOH$, etc. ammonia, nitrite oxidizers, NH_3, NO_2^-
		HETERO-	PROKARYOTES: "sulfur bacteria," S manganese oxidizers, Mn^{++} iron bacteria, Fe^{++}
	ORGANO-	AUTO-	PROKARYOTES: clostridia, etc., grown on CO_2 as sole source of carbon (H_2, $-CH_2$)
		HETERO-	PROKARYOTES (most) PROTOCTISTA (most) FUNGI PLANTS (achlorophyllous) ANIMALS

151

TABLE AI-2. THE DIVERSITY AND DISTRIBUTION OF BIOGENIC MINERALS IN EXTANT ORGANISMS. (From *On Biomineralization* by Heinz A. Lowenstam and Stephen Weiner. Copyright © 1989 by Oxford University Press, Inc. Reprinted by permission.)

	MINERAL	MAJOR	1	2	3	4	5	6	7	8	9	10	11	12	13	14	15	16	17	18	19	20	21	22	23	24	25	26	27	28	29
		KINGDOM						MONERA													PROTOCTISTA										
		PHYLUM NUMBER																													
CARBONATES	Calcite	Ca	■												■	■			■	■	■	■	■								
	Aragonite	Ca	■																	■	■	■	■	■							
	Vaterite	Ca																					■								
	Monohydrocalcite	Ca	■												■																
	Protodolomite	Ca, Mg	■																												
	Amorph hydrous carbonate	Ca	■																												
	Hydrocerussite	Pb												■																	
PHOSPHATES	Hydroxyapatite indet.[2]	Ca											■																		
	Octacalcium phosphate (OCP)	Ca					■						■																		
	Francolite	Ca												■																	
	Dahllite	Ca												■																	
	Ca₃Mg₃(PO₄)₄	Ca, Mg			■																										
	Whitlockite	Ca, Mg			■	■	■																								
	Struvite	Mg														■															
	Brushite	Ca																													
	Amorph pyrophosphate	Ca																													
	Amorph calcium phosphate (ACP) indet.	Ca																													
	ACP (dahllite precursor)[3]	Ca															■														
	ACP (brushite precursor)[3]	Ca																													
	ACP (whitlockite precursor)[3]	Ca																													
	ACP (francolite precursor)[3]	Ca																													
	Amorph Mg Ca phosphate	Mg, Ca																													
	Amorph hydrous Fe³⁺ phosphate	Fe, Ca	■					■	■			■																			
	K Na₃(Fe₆Mg₅)(PO₄)₃(OH₃)[3]	Fe																													
	Vivianite Fe₃²⁺(PO₄)₂·8H₂O	Fe									■																				
HALIDES	Fluorite	Ca																													
	Amorph fluorite	Ca																													
	Hieralite	K																													
SULFATES	Gypsum	Ca																						■	■						
	Celestite	Sr																							■	■					
	Barite	Ba																						■		■					
	Jarosite	K, Fe								■																					
SILICA	Opal	Si													■	■					■	■						■	■	■	■

		Element
IRON OXIDES	Magnetite	Fe
	Goethite	Fe
	Lepidocrocite	Fe
	Ferrihydrite	Fe
	Amorph iron oxide	Fe
	Amorph "ilmenite"	Fe, Ti
MANGANESE OXIDES	Todorokite	Mn
	Birnessite	Mn
SULFIDES	Pyrite	Fe
	Hydrotroilite	Fe
	Sphalerite	Zn
	Wurtzite	Zn
	Galena	Pb
	Greigite	Fe
	Mackinawite	Fe
METALS	Sulfur	–
CITRATE	Earlandite	Ca
OXALATES	Whewellite	Ca
	Weddelite	Ca
	Glushinskite	Mg
	$Mn_2C_2O_4 \cdot 2H_2O$	Mn
	$Cu\ C_2O_4 \cdot nH_2O$	Cu
	Ca oxalate indet.	Ca
OTHER ORGANIC CRYSTALS	Sodium urate	Na
	Uric acid	–
	Paraffin hydrocarbon	–
	Wax (long chain)	–
	Ca tartrate	Ca
	Ca malate	Ca

1   Taxonomic assignments are arbitrarily based on Margulis and Schwartz (1988)
2   Hydroxyapatite is often loosely used for apatite minerals that also contain carbonate and/or fluorine. We do not imply that the designations shown signify that these organisms form hydroxyapatite and not one of the other forms.
3   The term "precursor" refers to an amorphous phase which upon heating to 500°C converts to the designated crystalline phase.
4   Found in bacterial ferritin

TABLE AI-2. (*Continued*)

	MINERAL	MAJOR	FUNGI 30 31 32 33	PLANTAE 34 35 36 37 38 39 40 41 42 43	ANIMALIA 44 45 46 47 48 49 50 51 52 53 54 55 56 57
		KINGDOM / PHYLUM NUMBER			
CARBONATES	Calcite	Ca			
	Aragonite	Ca			
	Vaterite	Ca			
	Monohydrocalcite	Ca			
	Protodolomite	Ca, Mg			
	Amorph hydrous carbonate	Ca			
	Hydrocerussite	Pb			
PHOSPHATES	Hydroxyapatite indet.[2]	Ca			
	Octacalcium phosphate (OCP)	Ca			
	Francolite	Ca			
	Dahllite	Ca			
	$Ca_3Mg_3(PO_4)_4$	Ca, Mg			
	Whitlockite	Ca, Mg			
	Struvite	Mg			
	Brushite	Ca			
	Amorph pyrophosphate	Ca			
	Amorph calcium phosphate (ACP) indet.	Ca			
	ACP (dahllite precursor)[3]	Ca			
	ACP (brushite precursor)[3]	Ca			
	ACP (whitlockite precursor)[3]	Ca			
	ACP (francolite precursor)[3]	Ca			
	Amorph Mg Ca phosphate	Mg, Ca			
	Amorph hydrous Fe^{3+} phosphate	Fe, Ca			
	$K\,Na_3(Fe_{15}Mg_{25})\text{-}(PO_4)_3(OH_3)$[3]	Fe			
	Vivianite $Fe_3^{2+}(PO_4)_2 \cdot 8H_2O$	Fe			
HALIDES	Fluorite	Ca			
	Amorph fluorite	Ca			
	Hieratite	K			
SULFATES	Gypsum	Ca			
	Celestite	Sr			
	Barite	Ba			
	Jarosite	K, Fe			

154

Category	Mineral	Element
SILICA	Opal	Si
IRON OXIDES	Magnetite	Fe
	Goethite	Fe
	Lepidocrocite	Fe
	Ferrihydrite	Fe
	Amorph iron oxide	Fe
	Amorph "ilmenite"	Fe, Ti
MANGANESE OXIDES	Todorokite	Mn
	Birnessite	Mn
SULFIDES	Pyrite	Fe
	Hydrotroilite	Fe
	Sphalerite	Zn
	Wurtzite	Zn
	Galena	Pb
	Greigite	Fe
	Mackinawite	Fe
METALS	Sulfur	–
CITRATE	Earlandite	Ca
OXALATES	Whewellite	Ca
	Weddelite	Ca
	Glushinskite	Mg
	$Mn_2C_2O_4 \cdot 2H_2O$	Mn
	$Cu\,C_2O_4 \cdot nH_2O$	Cu
	Ca oxalate indet.	Ca
OTHER ORGANIC CRYSTALS	Sodium urate	Na
	Uric acid	–
	Paraffin hydrocarbon	–
	Wax (long chain)	–
	Ca tartrate	Ca
	Ca malate	Ca

1 Taxonomic assignments are arbitrarily based on Margulis and Schwartz (1988)
2 Hydroxyapatite is often loosely used for apatite minerals that also contain carbonate and/or fluorine. We do not imply that the designations shown signify that these organisms form hydroxyapatite and not one of the other forms.
3 The term "precursor" refers to an amorphous phase which upon heating to 500°C converts to the designated crystalline phase.
4 Found in bacterial ferritin

Phylum number	Kingdom	Phylum
1	Monera	Cyanobacteria
2		Pseudomonads
3		Actinobacteria
4		Fermenting bacteria
5		Omnibacteria
6		N_2-fixing aerobic bacteria
7		Aphragmabacteria
8		Aeroendospore
9		Chemoautotrophic bacteria
10		Thiopneutes
11		Micrococci
12		Undetermined
13	Protoctista	Myxomycota
14		Ciliophora
15		Rhizopoda
16		Foraminifera
17		Dinoflagellata
18		Zoomastigina
19		Haptophyta
20		Rhodophyta
21		Chlorophyta
22		Phaeophyta
23		Gamophyta
24		Actinopoda
25		Bacillariophyta
26		Xanthophyta
27		Pyrrhophyta
28		Chrysophyta

Phylum number	Kingdom	Phylum
29		Euglenophyta
30	Fungi	Ascomycota
31		Basidiomycota
32		Deuteromycota
33		Mycophycophyta
34		Zygomycota
35	Plantae	Bryophyta
36		Sphenophyta
37		Filicinophyta
38		Coniferophyta
39		Gnetophyta
40		Ginkgophyta
41		Angiospermophyta
42		Cycadophyta
43		Lycopodophyta
44	Animalia	Porifera
45		Cnidaria
46		Platyhelminthes
47		Nemertina
48		Ectoprocta
49		Brachiopoda
50		Annelida
51		Mollusca
52		Arthropoda
53		Sipuncula
54		Pogonophora
55		Echinodermata
56		Chordata

[1] For references, see Lowenstam and Weiner, 1989.

biosphere   The area at the surface of the Earth occupied or favorable for occupation by living organisms including parts of the lithosphere (q.v.), hydrosphere (q.v.), and the lower atmosphere (q.v.)

biota   All living organisms of an area; the sum of all living things on Earth

budding   Reproduction (q.v.) by outgrowth of a small protrusion (i.e., a bud) from a parent cell or body

carnivore   Predator that obtains its nutrients and energy by eating meat

CBWE   Corn Blight Watch Experiment

chaparral   Ecosystem type; vegetation consisting of thickets of dwarfed, drought-resistant, and often thorny shrubs and bushes, characteristic of much of the western United States

chemotrophy   Nutritional mode of organisms that obtain their energy from inorganic compounds (see Table AI–1, p. 151)

chert   A hard, extremely dense or compact, dull to semilustrous, crypto-crystalline (q.v.) sedimentary rock, consisting mostly of cryptocrystalline silica with lesser amounts of micro- or cryptocrystalline quartz and amorphous silica (opal); chert, which is the major rock type for fine preservation of microfossils, sometimes contains calcite, iron oxide, and the remains of silicious and other organisms. Chert occurs principally as nodules in limestones and dolomites, and less commonly as thick, layered deposits

chlorophyll   Green pigment responsible for absorption of visible light in photosynthetic organisms; all are magnesium chelated heterocyclic porphyrin derivatives (e.g., organic ring compounds) with isoprenoid side chains

chloroplast   Green plastid, a membrane-bounded photosynthetic organelle of eukaryotic cells containing chlorophylls a and b, and capable of oxygenic photosynthesis

chromatid   Half a chromosome in prophase and metaphase; becomes a chromosome in anaphase of mitosis

chromatin   The complex of nucleic acids (DNA and RNA) and proteins (histones and nonhistones) comprising chromosomes, the component fibril comprised of nucleosomes and spacers (limited to eukaryotes)

chromosome   Staining structures in the eukaryotic nucleus, composed of chromatin and carrying genetic information arranged in a linear sequence. Term should be restricted to eukaryotes, but is also used to refer to genophore (q.v.)

CITARS   Crop Identification Technology Assessment for Remote Sensing (Experiment)

coccolith   (a) Scales of certain haptophytes (haptomonads, prymnesiophytes), microscopic calcareous structural elements or buttonlike plates having many different shapes and averaging about 3 microns in diameter (some have diameters as large as 35 microns), constructed of minute calcite or aragonite crystals, and constituting the outer skeletal remains of a coccolithophore. Coccoliths are found in chalk and in deep-sea oozes of the temperate and tropical oceans, and were probably not common before the Jurassic. (b) Two shields connected by a central tube of a coccolithophorid

coccolithophore or **coccolithophorid**   Any of numerous, minute, mostly marine, planktonic biundulipodiated protists having brown pigment-bearing cells that at some phase of their life cycles are encased in a sheath of coccoliths (q.v.) to form a complex calcareous shell; the mastigote, haptonemate organism that produces coccoliths

colligative property   A property that depends on the number of particles present in a substance, rather than on the nature of the particles. Examples are osmotic pressure, the elevation of boiling point, depression of freezing point, or change in vapor pressure of solvents caused by dissolved substances

community   A unit in nature comprised of populations of organisms of different species living in the same place at the same time; microbial communities are those lacking significant populations of animals and plants

consumer   A heterotroph (q.v.) that derives its energy from living or freshly killed organisms or parts thereof. Primary consumers are herbivores (q.v.); higher-level consumers are carnivores (q.v.)

corrasion   A process of erosion (q.v.) whereby rocks and soil are mechanically removed or worn away by the abrasive action of solid materials moved along by wind, waves, running water, glaciers, or gravity (e.g., the wearing away of the bed and banks of a stream by the cutting, scraping, scratching, and scouring effects of a sediment load carried by the stream or the sawing and grinding action of sand, gravel, and boulders hurled by waves and currents against a shore)

cryptocrystalline   The texture of a rock consisting of or having crystals that are too small to be recognized and separately distinguished even under the ordinary microscope (although crystalline features may be shown by use of the electron microscope)

DBM   Drainage Basin Model

decomposer   Small detritivores (q.v.), usually bacteria (q.v.) and fungi (q.v.), that consume such substances as cellulose and nitrogenous waste products. Their metabolic processes release inorganic nutrients, which are then available for reuse by plants and other organisms

denitrification   An anaerobic respiratory process characteristic of facultative aerobic bacteria growing under oxygen-depleted conditions (denitrifying bacteria). The reduction of nitrate to nitrite or to gaseous products such as nitrogen, nitrous oxide, and nitric oxygen

detritivores Organisms that live on dead and discarded organic matter, include large scavengers, smaller animals such as earthworms and some insects, and decomposers (q.v.)

D-layer A layer in the ionosphere (q.v.), existing below 90 km above the Earth's surface (see Figure i–1, p. xi)

DNA Deoxyribonucleic acid; a long molecule composed of nucleotides, the linear order of which constitutes the genetic information of cells; capable of self-replication and of determining the nucleotide sequence in synthesis of complementary RNA (q.v.)

drainage basin A region or area bounded peripherally by a drainage divide and occupied by a drainage system; specifically, the whole area or entire tract of country that gathers water originating as precipitation and contributes it ultimately to a particular stream channel or system of channels, or to a lake, reservoir, or other body of water

ecosystem A visibly recognizable unit in nature, bounded by an ecotone (q.v.) and comprised of communities in which the biologically important chemical elements (C, N, S, P, and so on) are cycled; these elements cycle more rapidly within an ecosystem than between ecosystems. The verification of rates of intraecosystem cycling are under extensive study. A pond, forest,or chaparral are all examples of an ecosystem.

ecotone Ecosystem (q.v.) boundary, transition zone between one ecosystem and another (e.g., field or forest edge, shoreline, etc.)

E-layer A layer in the ionosphere (q.v.), existing at approximately 90 to 120 km above the Earth's surface (see Figure i–1, p. xi)

embryo Early developmental stage, characterizing only members of the animal or plant kingdoms; develops from a zygote (fertilized egg)

endoplasmic reticulum (ER) A system of membranes forming sheets and vesicles in the cytoplasm of most eukaryotes. In places the ER is continuous with the plasma membrane or the outer membrane of the nuclear envelope. If the outer surfaces of the ER membranes are coated with ribosomes, the ER is termed rough-surfaced; if ribosomes are sparse or absent, the ER is called smooth-surfaced. Absent in bacteria

EOS Earth Observing System (Satellite System)

epidermis In plants and animals, the outermost layer of cells

ERBE Earth Radiation Budget Experiment

erosion The general process or the group of processes whereby the earthy and rocky materials of the Earth's crust are loosened, dissolved, or worn away and simultaneously removed from one place to another by natural agencies that include weathering (q.v.), solution (q.v.), corrasion (q.v.), and transportation (q.v.), but usually exclude mass-wasting (q.v.); specifically, the mechanical destruction of the land and the removal of material (such as soil) by running water (including rainfall), waves and currents, moving ice, or wind

ERTS Earth Resources Technology Satellite

eukaryote Cell, or organism comprised of cells, having membrane-bounded nuclei. Membrane-bounded organelles (mitochondria and plastids) and several chromosomes are typical of eukaryotes. Unit of structure of protoctists, animals, plants, and fungi

evolution   (a) The term referring to the phenomenon of common ancestry, appearance, diversification, change, and extinction of organisms throughout the history of the Earth. (b) The mechanism by which approximately 10 million extant species of organisms appeared. This mechanism involves reproduction with high heritability (ultimately based on the transmission of replicating nucleic acids—DNA and RNA—comprised of long complementary base sequences and organized into genes) and occasional heritable change (by mutation in the broad sense: changes in the quality and quantity of genes, genophore and chromosomal rearrangements as well as heritable associations of parts and whole organisms). The fact that biotic potential is not reached because it is limited by scarcities of water, food, living space, predation, competition, and other factors is the fact of natural selection that leads to differential reproduction of heritable variants and ultimately to new populations, species, and higher taxa from a common ancestor.

exobase The lower boundary of the exosphere (q.v.), located at about 500 km above the Earth's surface (see Figure i–1, p. xi)

exosphere The most remote, isothermal region of the atmosphere extending from about 500 km above the Earth's surface and merging with interplanetary space (see Figure i–1, p. xi)

F-layer A layer in the ionosphere (q.v.), which is a daytime feature centered at about 150 km above the Earth's surface (see Figure i–1, p. xi)

F–2 layer A layer in the ionosphere (q.v.), centered at about 300 km above the Earth's surface (see Figure i–1, p. xi)

FAO Food and Agricultural Organization

fission (a) Binary fission (q.v.); (b) nuclear fission (q.v.)

FOCIS Forest Classification and Inventory System (Experiment)

foraminifer or **foraminiferan** Protoctist belonging to the phylum Granuloreticulosa. Foraminifera are characterized by the presence of a test (shell) composed of agglutinated particles or of secreted calcite (only rarely of silica or aragonite). These multinucleate protists are commonly found in marine to brackish environments from the Cambrian to the present. Pre-Cambrian forms and some freshwater forms are also known; short name: forams

fungus   Member of the kingdom Fungi. Haploid, dikaryotic, occasionally diploid organisms typically displaying zygotic meiosis. Osmotrophic heterotrophic eukaryotes that reproduce from conidia, zygo-, asco-, or basidiospores. Amastigote at all life cycle stages, the kingdom includes molds, mushrooms, puffballs, shelf fungi, and lichens

G²DAS   Geo-referenced Global Data and Analysis System

gaia   (a) Mother Earth (ancient Greek), the goddess of the Earth. (b) The Gaia hypothesis states that the surface of the Earth is regulated by the activities of life. Specifically, the Earth's atmosphere is maintained far from chemical equilibrium with respect to its composition of reactive gases, oxidation-reduction state, alkalinity-acidity, albedo, and temperature. This environmental maintenance is affected by the growth and metabolic activities of the sum of the organisms (i.e., the biota, q.v.). The Gaia hypothesis implies that the surface conditions on Earth would revert to those interpolated for a planet between Mars and Venus were life to be eliminated.

gedanken experiment   Thought experiment

genophore   In prokaryotes, the circular DNA molecule containing a set of the genetic instructions for the cell

geognosy   Knowledge of the Earth; science of planet Earth

GERO   Global Environmental Research Organization (Universities of California, Venice, and Padua) (See Appendix II, pp. 171-184)

gigamole 10^9 moles (q.v.)

grassland   Ecosystem type: area in which grasses constitute the dominant vegetation, as in the prairies or plains region of the United States, the pampas of South America, and the steppes of Russia

guard cells   Specialized epidermal cells surrounding a pore, or stoma (q.v.), in a leaf or green stem; changes in turgor of a pair of guard cells cause opening and closing of the pore

GSFC   Goddard Space Flight Center (Greenbelt, Maryland, United States)

HCMM   Heat Capacity Mapping Mission (Satellite)

herbivore   A consumer (q.v.) that eats plants or other photosynthetic organisms to obtain its food and energy

heterotrophy   Nutritional mode of organisms that gain both carbon and energy from organic compounds (ultimately produced by autotrophs) (see Table AI–1, p. 151)

HIRIS   High Resolution Imaging Spectrometer

histones Class of positively charged chromosomal proteins that bind to DNA and comprise chromatin; they are rich in lysine and arginine amino acid residues

HMRR High Resolution Multifrequency Microwave Radiometer

homeostasis When a regulatory system (e.g., physiological) actively maintains specified variables at relatively constant levels (i.e., set points) in spite of perturbing influences

homeorrhesis When a regulatory system (e.g., physiological) actively maintains specified variables at relatively constant levels even though the variables change through time (i.e., moving set points or operating points) in spite of perturbing influences

HRV High Resolution Visible Instruments

hydroid Animal. Member of Phylum Cnidaria (Coelenterata). Any one of a group of hydrozoans belonging to the order Hydroida, among which the polyploid (usually colonial) generation is dominant and whose skeleton is commonly composed of a hornlike material

hydrosphere The Earth's waters, as distinguished from its rocks (lithosphere, q.v.), its living things (biota, q.v.), and its air (atmosphere, q.v.); includes surface waters in liquid form (for example, oceans, rivers, and lakes) and in solid form (for example, snow, ice, and glaciers); can also include water vapor, clouds, and all forms of precipitation still in the atmosphere

hypha One of the individual tubular filaments or threads that make up the mycelium (q.v.) of a fungus (pl.: hyphae)

ICSU International Council of Scientific Unions

IFOV Instrument Field of View

ionosphere Part of the Earth's atmosphere; this region consists of a series of constantly changing layers of heavily ionized molecules, termed the D-layer (q.v.), E-layer (q.v.), F-1 layer (q.v.), and F-2 layer (q.v.), which are superimposed upon the other, nonionized atmospheric regions (see Figure i–1, p. xi)

LACIE Large Area Crop Inventory Experiment

LAI Leaf Area Index

Landsat Land satellite (formerly ERTS, q.v.)

LARS Laboratory for applications of Agricultural Remote Sensing

LASA Lidar Atmospheric Sounder and Altimeter

lipid One of a class of organic compounds, soluble in organic but not aqueous solvents; the class includes fats, waxes, steroids, phospholipids, carotenoids, and xanthophylls

lithosphere The solid portion or crust of the surface of the Earth, comprised of a number of mobile plates. The plates from which the continents emerge are composed of primarily basaltic, granitic, and various sedimentary rocks. The lithosphere is distinguished from the atmosphere (q.v.) and the hydrosphere (q.v.), as well as from the Earth's interior—its mantle (including the upper plastic aesthenosphere) and core

lithotrophy Chemoautotrophy. Metabolic mode in which oxidation of inorganic compounds provides the source of energy for metabolism (see Table AI–1, p. 151)

lysosome A membrane-enclosed intracellular vesicle, the primary cell component for intracellular digestion in eukaryotes. Lysosomes are known to contain at least 50 acid hydrolases, including phosphatases, glycosidases, proteases, sulfatases, lipases, and nucleases. Collectively, lysosomes can hydrolyze all classes of macromolecules.

magnetosphere The region around the Earth to which the Earth's magnetic field is confined, due to interaction between the solar wind and the geomagnetic field. On the sunlit side, the magnetosphere is approximately hemispherical, with a radius of about ten Earth radii under quiet conditions; it may be compressed to about six Earth radii by magnetic storms. On the side opposite the sunlit side, the magnetosphere extends in a "tail" of several hundred Earth radii.

mass wasting A general term for the dislodgement and downslope transport of soil and rock material under the direct application of gravitational body stresses. In contrast to other erosion (q.v.) processes, the debris removed by mass wasting processes is not carried within, on, or under another medium possessing contrasting properties.

meiosis A process of eukaryotes involving one or two nuclear divisions, in which the number of chromosomes (q.v.) is reduced by half. The reciprocal process of fertilization in which diploid or 2N cells are reduced to haploid or 1N cells (such as eggs, sperm, spores)

mesopause The upper boundary of the atmosphere (q.v.) (see Figure i–1, p. xi)

mesosphere (a) The portion of Earth's atmosphere from about 50 to 80 km in which the temperature, with increasing height, at first increases to a maximum of around 280 K (45° F) and then decreases to around 180 K (−135° F) or less, depending on the latitude and the season. Its lower boundary is the stratopause (q.v.) and the upper boundary is the mesopause (q.v.). (b) A similar region in the atmosphere of other planets (see Figure i–1, p. xi)

metabolism The sum of all enzyme-mediated chemical conversion pathways characteristic of all autopoietic entities, cells, and organisms (e.g., the means by which energy and organic compounds are made available for use by organisms)

microbe Microscopic organism, includes all bacteria and most protoctists and fungi

microbial mat A community (q.v.) of microorganisms forming a flat, cohesive structure and laminated cherts (q.v.); living precursors of stromatolites

mitochondrion An organelle in which the chemical energy in reduced organic compounds (food molecules) is transferred to ATP molecules by oxygen-respiring respiration

mitosis Cell division process characteristic of most eukaryotes. Nuclear division in which attached chromatids (chromosomes composed of condensed chromatin in which DNA and protein have already duplicated) move to the equatorial plane of the nucleus, separate at their kinetochores, and form two separate, equal groups of chromatin

mixotrophy Nutritional mode of bacteria and protoctists characteristic of organisms nourished by both autotrophic and heterotrophic mechanisms

MODIS Moderate Resolution Imaging Spectrometer

mole The basic SI unit of amount of substance equal to the number of entities as there are atoms in 0.012 kilogram of carbon–12. The entities may be atoms, molecules, ions, electrons, or similar elementary units. One mole of any substance contains N entities, where N is the Avagadro's number (6.022×10^{23}). One mole of a compound is equivalent to M grams, where M is the molecular weight

MSS Multispectral Scanner

mycelium The visible mass of a fungus; "fuzz" composed of intertwined hyphae (q.v.)

NASA National Aeronautics and Space Administration

natural selection See **evolution**

nitrification The oxidation of ammonia (in solution as a salt) to nitrite, often followed by the oxidation of nitrite to nitrate. An aerobic respiratory process characteristic of chemolithotrophs (nitrosofying and nitrifying bacteria). This process enriches the nitrogen content of soil.

nitrogen fixation Metabolic process characteristic of some bacteria: incorporation of atmospheric nitrogen (N_2) into organic nitrogen compounds; requires nitrogenase (q.v.)

nitrogenase Enzyme complex containing iron and molybdenum; reduces atmospheric nitrogen to organic nitrogen compounds

NOAA National Oceanic and Atmospheric Association

NPP Net Primary Production

nuclear fission A transformation of atomic nuclei characterized by the splitting of a nucleus into at least two other nuclei and the release of amounts of energy far greater than those generated by conventional chemical reactions

nucleotide   Single unit of nucleic acid; composed of an organic nitrogenous base, linked to deoxyribose or ribose sugar, and phosphate

nucleus   Characteristic intracellular organelle of eukaryotes: membrane-bounded structure that contains most of a cell's genetic information in the form of chromatin (q.v.)

organelle   "Little organ"; a distinguishable part of a cell intracellular structure composed of an organized complex of macromolecules and small molecules (e.g., nucleus, mitochondrion, plastid, ribosome, and mesosome)

organotrophy   Nutritional mode of organisms that obtain their electrons from organic compounds

otiose   (a) Unemployed, at one's ease, indolent; (b) ineffective, futile, sterile; (c) useless, superfluous

ozone layer   A region in the upper stratosphere (q.v.) containing high concentrations of ozone (gaseous triplet oxygen, O_3), which protects organisms at the Earth's surface from harmful solar radiation (see Figure i–1, p. xi)

parthenogenesis   The development of an individual from an egg without fertilization

Phanerozoic Eon   The modern eon of Earth history, beginning 570 million years ago with the appearance of Cambrian hard-bodied fossil forms and lasting until the present

photogrammetry   The science or art of obtaining reliable measurements by means of photographs; specifically, mapmaking and surveying with the aid of aerial and terrestrial photographs

photo-interpretation   The science of identifying and describing objects. imaged in a photograph, such as deducing the topographic significance or the geologic structure of landforms from an aerial photograph

photosynthesis   Metabolic process involving the production of organic compounds from carbon dioxide and a hydrogen donor (e.g., hydrogen sulfide or water) by using light energy captured by chlorophyll (q.v.)

phylogeny   The relationships of groups of organisms as reflected by their evolutionary history

plant   Member of the Kingdom Plantae. Eukaryotes comprised of cells containing plastids (e.g., chloroplasts, their precursors or derivatives) that develop from nonblastular embryos. The Kingdom includes mosses, ferns, conifers, and flowering plants.

plastid Cytoplasmic membrane-bounded organelle of plants and algae, either photosynthetic (e.g., chloroplast, rhodoplast) or a nonphotosynthetic derivative (e.g., amyloplast)

population A group of organisms belonging to the same species and living in the same place at the same time

producer An autotrophic (q.v.) organism, usually a photosynthesizer (q.v.), that contributes to the net primary productivity of a community (q.v.)

prokaryote Bacterial cell or organism, lacking a membrane-bounded nucleus and membrane-bounded organelles. All are members of the kingdom Monera (Prokaryotae)

protein Macromolecule consisting of long chain of amino acids linked by peptide bonds. Some are enzymes, which hasten chemical reactions in living organisms; others play a structural role (e.g., tubulin, actin, and myosin).

protist Microscopic, usually single-celled, member of the kingdom Protoctista; an informal name for heterotrophic (protozoa) or autotrophic (algae) eukaryotic microorganisms

protoctist Member of the kingdom Protoctista. Eukaryotic, heterotrophic, and autotrophic microorganisms and their larger descendants, exclusive of animals, plants, and fungi. None form embryos. The kingdom includes diatoms, dinoflagellates, brown seaweeds and other algae, ciliates, amoebas, malarial parasites, slime molds, slime nets, and many other groups.

radiation flux Amount of radiation impinging on a given surface per unit time

respiration Metabolic process that uses electron transport chains and involves the breakdown of organic compounds in the release of energy; the terminal electron acceptor is inorganic and may be oxygen or (in anaerobic organisms) nitrate, sulfate, or nitrite

ribosome A spherical organelle in all cells composed of protein and ribonucleic acid; the site of protein synthesis

riparian Pertaining to or situated on the bank of a body of water, especially of a water course such as a river (e.g., "riparian land" situated along or abutting upon a stream bank)

rivershed The drainage basin (q.v.) of a river

RNA Ribonucleic acid; a molecule composed of a linear sequence of nucleotides which can store genetic information; a component of ribosomes, it takes part in protein synthesis

saprobe Organism that excretes extracellular digestive enzymes and absorbs dead organic matter

saprophyte saprobe (q.v.)

SAM   Sensing with Active Microwaves

SAR   Synthetic Aperture Radar

Seasat   Sea satellite

SEM   Scanning Electron Microscope

sexual reproduction   Reproduction leading to individual offspring having more than a single parent

SISP   Surface Imaging and Sounding Package; Sensor subsystem on EOS (q.v.) platform

solution   A process of chemical weathering by which rock material passes into solution (e.g., the dissolution and removal of the calcium carbonate in limestone or chalk by carbonic acid derived from rainwater containing carbon dioxide acquired during its passage through the atmosphere)

spore   Small or microscopic propagative unit capable of development into a mature or active organism; often desiccation- and heat-resistant

SPOT   System Probaterie de le Observatoire Terre (Satellite System)

stoma, pl. stomata   A minute opening bordered by guard cells (q.v.) in the epidermis (q.v.) of leaves and stems through which gases pass

stratopause   The upper boundary of the stratosphere (q.v.) (see Figure i–1, p. xi)

stratosphere   (a) An upper portion of a planetary atmosphere, above the troposphere and below the mesosphere, characterized by relatively uniform temperatures and horizontal winds. On Earth, its lower limit varies from about 8 to 20 km; its upper limit is at around 45 km. The temperature in this region is around −75° C. The base of the stratosphere makes an upper limit to the general turbulence and convective activity of the troposphere; thus, air motion within the stratosphere is largely horizontal (jet stream) (see Figure i–1, p. xi)

stromatolite   Laminated carbonate or silicate rocks; organo-sedimentary structures produced by growth, metabolism, trapping, binding, and/or precipitating of sediment by communities of microorganisms, principally cyanobacteria. Lithified or fossil form of microbial mat communities

TEM   Transmission Electron Microscope

TEM   Terrestrial Ecosystem Model

terraforming   The modification of a planetary surface so that it supports life

TGM   Trace Gas Model

thermosphere   An atmospheric layer extending from 80 km to roughly 500 km above the Earth, in which temperature increases; the temperature of the thermosphere is highly variable because it is dependent on solar activity and changes with time of day and, to some extent, with latitude (see Figure i–1, p. xi)

TIROS   Television and Infra-Red Observation Satellite

transpiration   In plants, the loss of water vapor from the stomata (q.v.)

transportation   A phase of sedimentation concerned with the actual movement, shifting, or carrying away by natural agents (such as flowing water, ice, wind, or gravity) of sediment or of any loose, broken, or weathered material, either as solid particles or in solution, from one place to another (over a short or long distance) on or near the Earth's surface; e.g., the drifting of sand along a seashore under the influence of currents, the creeping movement of rocks on a glacier, and the conveyance of mud and dissolved salts by a stream

tropopause   The upper boundary of the troposphere (q.v.) (see Figure i–1, p. xi)

troposphere   The lowest layer of a planetary atmosphere in which the temperature decreases steadily with increased altitude, extending from the Earth's surface to its upper boundary, the tropopause (q.v.), at a height of about 10 to 20 km, depending on the latitude and time of year. Turbulence is greatest in this region, and most of the visible phenomena associated with the weather occurs here (for example, cloud formation) (see Figure i–1, p. xi)

TM   Thematic Mapper

UARS   Upper Atmosphere Research Satellite

USDA   United States Department of Agriculture

USGS   United States Geological Survey

virus   A large group of infectious agents ranging from 10 to 250 nanometers, in diameter. Generally composed of a protein layer surrounding a nucleic acid core and capable of infecting animals, plants, or bacteria; nonautopoietic entities dependent entirely on living cells for reproduction and metabolism

watershed   (a) The line, ridge, or summit of high ground separating two drainage basins. (b) The region drained by, or contributing water to, a stream, lake, or other body of water

weathering   The destructive process or group of processes constituting that part of erosion (q.v.) hereby earthy and rocky materials on exposure to atmospheric agents at or near the Earth's surface are changed in character (color, texture, composition, firmness, or form), with little or no transport of the loosened or altered material; specifically, the physical disintegration and chemical decomposition of rock that produce an in situ mantle of waste and prepare sediments for transportation. Most weathering occurs at the surface, but it may take place at considerable depths, as in well-jointed rocks that permit easy penetration of atmospheric oxygen and circulating suface waters

wetland   Area where water is the primary factor controlling the environment and the associated plant and animal life; transitional habitats occur between upland and aquatic environments where the water table is at or near the surface of the land, or where the land is covered by shallow water (up to 6 feet deep)

The composition of this glossary was aided by the following sources:

Curtis, H. 1983. *Biology*, 4th ed. Worth Publishers, Inc., New York.

Daintith, J. 1976. *A Dictionary of Physical Sciences*. Rowman and Allanheld Publishers, New Jersey.

Gary, M., R. McAfee, Jr., and C.L. Wolf, eds. 1972. *Glossary of Geology*. American Geological Institute, Washington, DC.

King, R.C. and W.D. Stansfield. 1985. *A Dictionary of Genetics*, 3d ed. Oxford University Press, New York. c1968.

Lowenstam, H. and S. Weiner 1989. *On Biomineralization*. Oxford University Press, New York.

Margulis, L. 1982. *Early Life*. Science Books International, Inc., Boston.

Margulis, L. and D. Sagan. 1986. *Origins of Sex*. Yale University Press, New Haven and London.

Margulis, L. and Schwartz, K.V. 1988 *Five Kingdoms: An illustrated guide to the phyla of life on Earth*, 2nd edition. W.H. Freeman Co., New York.

Parker, S., ed. 1984. *McGraw-Hill Dictionary of Scientific and Technical Terms*, 3d ed. McGraw-Hill Book Co., New York.

Steen, E.B. 1971. *Dictionary of Biology*. Barnes and Noble, New York.

Tver, D.F. 1979. *Dictionary of Astronomy, Space and Atmospheric Phenomena*. Van Nostrand and Reinhold Co., New York.

II GLOBAL ENVIRONMENTAL RESEARCH ORGANIZATION (GERO)[1]

Modern civilization is changing the very planet on which we live. What once seemed immutable—the air we breathe, the oceans we sail, the ground we walk—we have changed. We do not know which changes will prove harmful and which will prove beneficial. We lack the scientific understanding to predict effects of far-reaching changes. International environmental changes are without historical precedent and require an unprecedented solution: an international center where research is actively pursued to seek solutions at all levels of our environmental problems.

The University of California, Santa Barbara; the University of Venice; the University of Padua; and the City of Venice have now joined together to develop rapid, efficient, and inexpensive methods to study, understand, and predict changes in our environment. GERO, the Global Environmental Research Organization, has been established to accomplish this goal.

GLOBAL

A mission of unprecedented scope challenges GERO: the creation of a new global science, the science of the biosphere. The biosphere, our plan-

---

[1] The information contained in this section was taken from a beautiful color pamphlet entitled "Global Environmental Research Organization." A limited number of these pamphlets may be available from Environmental Studies, University of California at Santa Barbara, Santa Barbara, CA 93106 USA.

etary life-support system, extends from the depths of the oceans to the upper reaches of the atmosphere. This new science is needed to understand, predict, and mitigate such changes as

- the increased acid rain on lakes, forests, agriculture, and urban areas,
- the widespread climate changes leading to drought and famine in Africa,
- the intensification of the Earth's "greenhouse effect" resulting from human-induced changes in our atmosphere,
- the effects of large-scale deforestation and subsequent creation of new desert areas.

To accomplish such truly ambitious goals, we must establish international cooperation in scientific research, data gathering, and applications of new knowledge. We must work to improve communication between nations and encourage a free flow of information.

GERO, with its multinational leadership and its research facility in Venice, Italy, will be uniquely able to guide this effort.

ENVIRONMENTAL

GERO's emphasis is on our planet's life-support system, the biosphere, including the Earth's large-scale cycles of water, chemicals, and energy flow. This requires the study of very large regions of the Earth, for which conventional techniques are not sufficient. Computers and satellite images play crucial roles, as illustrated here. GERO research will include

- the scientific nature of complex ecological systems,
- identification, research, and assessment of environmental problems and issues,
- technological opportunities for the solution to global environmental problems,
- economic implications of environmental issues.

Specific topics will reflect the goals of GERO as well as the interests of the visiting scientists and the needs of their home countries.

RESEARCH

One of the key tools that makes the study of the biosphere possible is remote sensing. GERO will use modern technology to interpret images obtained from satellites and aircraft.

But obtaining images is only the beginning. The images must be combined with other information, tested for accuracy and reliability, put into a form usable with ground information, and stored efficiently for easy retrieval.

ORGANIZATION

The international emphasis of the Global Environmental Research Organization is reflected in its two research centers to be located at the University of California, Santa Barbara and in Venice. Independently administered, the two centers share approaches, research projects, and goals.

SANTA BARBARA

As an integral part of the University of California, GERO will draw on the strengths and resources of one of the world's greatest institutions of research and scholarly studies. The Santa Barbara campus is a well-known center of study for some of GERO's most important topics: environmental sciences, remote-sensing applications, geography, computer science, and environmental policy and economics.

UC Santa Barbara is located 90 miles north of Los Angeles on an 815-acre promontory overlooking the Pacific Ocean. The 775 faculty members work together with a student population of approximately 16,000.

VENICE

Venice's 1000-year history as an international center for commerce, culture, and communication makes it an ideal location as a world center for the study of the biosphere. With its attractive locale and strong tradition of interest in international culture and civilization, the city is well situated for access from Europe, Africa, and Asia. The Venice center will be affiliated with two major Italian universities: the University of Venice and the University of Padua.

AGENDA FOR A NEW GLOBAL SCIENCE: SAFEGUARDING EARTH'S FUTURE[2]

- Our environmental problems, once simply local, now affect our planet at every scale—local, national, international, regional, and global.
- Current research on global issues lacks integrating concepts, interdisciplinary approaches, and statistical validity.

---

[2]The information contained in this section was taken from a pamphlet entitled "Agenda for a New Global Science: Safeguarding Earth's Future." A limited number of these pamphlets may be available from Environmental Studies, University of California at Santa Barbara, Santa Barbara, CA 93106 USA.

- We have powerful new tools for global research, such as satellite remote sensing, computer data handling, and precise environmental chemistry, but we lack the concepts that integrate these techniques into science.
- We lack a place for people to work together and use these techniques. There is a limit to what we can do with our present locally and regionally focused research efforts and concepts.
- To safeguard the future of Earth's life support system, we need a new science and a new way to conduct that science. We need a new global research institute that focuses on the biosphere—the global system that includes and sustains life.

GERO: THE NEW GLOBAL RESEARCH ORGANIZATION

GERO is a research institute where scientists from different disciplines can come together to study global problems using the best of modern technology. GERO research takes a problem-oriented approach to large-scale environmental issues, both regional and medium-term (1 to 10 years) and global and long-term (10 to 15 years or longer). These issues include such problems as acid rain, depletion of the ozone layer, climatic change, desertification, and the extinction of species.

Other international organizations that deal with environmental problems—such as the various United Nations agencies, the World Resources Institute, and SCOPE (the Scientific Committee on Problems of the Environment)—play a different role. Most act primarily as think tanks, integrators of information, or coordinators of dispersed research efforts. These organizations are essential, but they need the new information and new concepts that GERO can provide. GERO will work closely with such organizations.

GERO RESEARCH EMPHASIZES LIFE

The remarkable feature of our planet is life and its persistence for more than 3 billion years. Today there are between 3 and 10 million species on Earth. The focus of GERO is on life and the factors necessary to sustain it at the global level. The more we understand about the biosphere the more it becomes possible for us to take positive actions to support it—to provide for sustainable uses of natural resources, to protect essential parts of the system, and to enhance the quality of our environment.

GERO IS AN INTERNATIONAL ORGANIZATION

GERO was founded in 1985 as a cooperative activity by the University of California and the Universities of Padua and Venice in Italy. Research facilities are being developed at the University of California at Santa Barbara and at an affiliated center in Italy. Eventual plans call for a central research institute with ancillary centers at geographically representative locations throughout the world. The organization will bring together investigators from around the world to work on major environmental issues.

GERO IS A SCIENTIFIC ORGANIZATION

GERO is first and foremost a research institute concerned with basic scientific inquiry; it is not designed to set policy. However, understanding our environment in global terms is essential to the design of social, economic, and technological policies that will shape the future of the biosphere.

The primary mission of GERO is the creation of a new scientific discipline, the science of the biosphere. Research on the biosphere can be divided into five major themes:

- Sustaining Life on Earth
- Energy Flow through the Biosphere
- Cycling of Matter through the Biosphere
- Theory of the Biosphere
- Current State of the Biosphere.

All aspects of the biosphere interact—life affects energy flow and chemical cycles; changes in the flow of energy affect chemical cycles; changes in chemical cycles affect life. The key to understanding the biosphere is to recognize, study, and formulate concepts about these interactions. GERO offers the unique facilities and coordination required to develop this new science.

SUSTAINING LIFE ON EARTH

Sustaining life on Earth is the central topic of GERO. It is the most important issue to us, since we are alive and our survival requires the persistence of both our environment and many of our fellow creatures. The two qualities that distinguish life—self-reproduction and evolution—lead to the great capacity of living things to grow and adapt to changes in the environment.

To understand how life has been sustained on Earth and what will be required in the future, we need to know how each of these factors affects the persistence of life:

- biological diversity,
- the importance of certain species,
- disturbances and changes in the environment, and
- the interplay between life and its environment.

Human activities are rapidly altering the number of species and the abundance of organisms. We greatly increase the abundance of some species, like crops, which occupy huge areas in monotonous uniformity; we cause the extinction of other species. There are many alarms being sounded about the possible consequences of these actions, but there is too little real understanding of the long- or even the short-term effects of these actions.

EXAMPLES OF FUNDAMENTAL SCIENTIFIC QUESTIONS

Can We Bottle Life?

Some believe that we can sustain life just by closing up a small number of species in a container and providing a source of light and a sink for heat emitted from this system. But the longest anyone has maintained such a system is less than 20 years. Just how big and how complicated must an environment be to sustain life, and how many species must there be?

Why So Many Species?

The vast number of species on Earth raises several questions: Does this large number of species tend to increase the persistence of life? Does life tend to increase the chances of its own persistence by modifying the environment in ways that are beneficial to life? What are the long-term consequences of biological modifications of the environment?

The Impact of New Life Forms

During the history of life on Earth there have been a small number of major biological innovations—the rise of entire new life forms that are able to exploit new opportunities in the environment. How have these innovations affected the environment and the persistence of life?

The Impact of Environmental Changes

Our environment undergoes changes at many scales of time and space, including occasional global perturbations. Many forms of life are adapted to change and require changes of certain kinds in order to survive. What are the connections between environmental change and the persistence of life? What role will the changes induced by human activities play in the persistence of life?

EXAMPLES OF APPLIED QUESTIONS

Species Extinction—What Are the Risks?

The number of species on Earth is rapidly decreasing. Will this decrease the stability of life on Earth as a whole? Out of the millions of species are there certain ones that must survive for the rest of life to continue? If so, which ones are they? How can we tell which species are most important to save from extinction? How do we determine which species are essential to our existence? How can we decrease the chance of extinction of major groups of organisms?

The Endangered Life of the Forests

At the present rates of clearance, little of the planet's original tropical forests will remain by the end of the century. How can we prevent extinctions of tropical forest species in spite of the loss of forests? What areas must be maintained to avoid a catastrophic loss of species?

Is Re-Greening a Real Option?

We are removing life from large areas of the planet. Some believe that this is justified because, they argue, it is simple to return such areas to their former conditions merely by abandoning them and letting nature take its course. But under what conditions can life return to an area? And when are we willing to allow time for this recovery to take place?

ENERGY FLOW THROUGH THE BIOSPHERE

Life cannot exist without a flow of energy. Earth's energy is received from the sun, used by living things, and given off as heat. Heat energy is lost from Earth into space. Through the flow of energy, life is connected to the cosmos. The sun not only fuels life; it fuels the atmosphere and the oceans. Changes in the way energy flows through Earth's atmosphere and oceans affect life.

Climate dynamics are the result of interactions among the atmosphere, oceans, and land surface. This is an area of active dynamic research in which better and better computer models are being used to project climatic changes. A major stumbling block to advances in this field is lack of knowledge about the effects of life on climate. Future advancement requires integrating information from various disciplines concerned with the global environment, an integration that GERO can provide.

EXAMPLES OF FUNDAMENTAL SCIENTIFIC QUESTIONS

The Vital Link Between Life and Climate

How do large areas of land vegetation affect global climate? It is an old idea that "forests make rain," meaning that rainfall is higher directly over and downwind from forests. Some evidence suggests that forests may affect climate over very large areas by increasing the amount of water evaporated, decreasing water runoff, and changing the reflection and absorption of sunlight. The extent of these effects, however, is not clear.

Ocean Currents and Life

Energy received from the sun drives the currents in the oceans. If the sunlight reaching the ocean surface varies, or if climate changes, then the ocean currents can change. This in turn can affect the distribution and abundance of life in many parts of the ocean.

EXAMPLES OF APPLIED QUESTIONS

The Advance of the Deserts

It has been suggested that the spread of deserts is a self-enhancing process; as more vegetation is lost, the climate downwind becomes drier, leading to a further spread of deserts. Such a process may be increasing the deserts in Africa, which has contributed to recent catastrophic famine on that continent. If such a process exists, what can be done to break the cycle and slow or reverse the on-going loss of productive land to desertification?

The Retreat of the Forests

Logging and other human activities are resulting in large-scale loss of forests. How is global climate affected by deforestation over a large region? Does deforestation change the climate in a way that tends to increase the loss of forests in the future? Does it decrease the ability of forests to recover? Can reforestation over large areas affect weather and climate? What are some economically feasible solutions to deforestation and undesirable climatic changes?

The Multiple Effects of Human Pollution

The burning of fossil fuels is adding carbon dioxide to the atmosphere, which has the potential to warm the climate. Humans also add dust to the atmosphere, which can cool the climate. Will the sum of human activities result in a significant climate change? How will climate change affect agriculture and forestry?

CYCLING OF MATTER THROUGH THE BIOSPHERE

All life requires certain chemical elements, and these must be supplied continually in the right ratios and amounts. Too much or too little of any essential element can have undesirable effects. Chemical elements cycle throughout the entire biosphere. Human activities are changing these global chemical cycles.

Most global chemical cycles are poorly understood. We still need to improve our understanding of chemical pathways, the amounts of specific chemicals stored in various portions of the biosphere, and the rates of flow of these chemicals. We also need to learn what factors control the storage and flow of the chemical elements required for life.

EXAMPLES OF FUNDAMENTAL SCIENTIFIC QUESTIONS

The Nitrogen Cycle

Nitrogen is required for all life; at the same time, life affects the nitrogen cycle. How would the global cycle of nitrogen be affected by the loss of a large area of a major ecosystem? How large would this change have to be in one kind of ecosystem before it would affect the nitrogen supply to another kind?

The Sulfur Cycle

Sulfur compounds are reduced (converted from complex organic compounds to inorganic sulfur) in salt marshes. If we destroy 90 percent of the salt marshes of the world, what does this imply for the sulfur cycle?

Would it affect the supply of sulfur to near-shore marine ecosystems? What impacts could this change have on local economies, including fisheries and tourism?

EXAMPLES OF APPLIED QUESTIONS

The Ozone Question

The ozone layer in the atmosphere protects us and other living things from harmful ultraviolet light rays. However, there is evidence that the ozone layer has decreased in recent decades. Some believe that this decrease is due to the use of certain chemicals in aerosol spray cans. But many factors affect the ozone layer. Some chemical compounds produced by life in natural areas—such as salt marshes, the ocean, and forests—affect chemical reactions in the atmosphere, which in turn affect the ozone layer. We must improve our understanding of the chemistry of the bio-sphere if we are to answer the ozone question: Will the ozone layer decrease to the point that it significantly increases cancer rates in people and death rates in many living things?

Heavy Metal Pollution

How is the production of fish being affected by the introduction of heavy metals into the ocean? What are economically feasible ways to contain these heavy metals in local areas to reduce their effects on fisheries?

Acid Rain

The burning of fuels adds acids to the air. The acid has definite effects on lakes and may be killing trees over large areas of Europe and North America. This loss of trees can increase soil erosion and the transfer of chemical elements from the land to the sea. It also changes the evaporation of water from the land and the absorption and reflection of sunlight; these changes, in turn, affect climate. Thus, burning fuels may change the dis-tribution of life, which can have a large effect on chemical cycles and energy flow.

Agricultural Pollutants

Agricultural fertilizers and pesticides add chemicals to the soil. Some of these are carried by flowing waters to the seas and oceans, where they can reach levels that cause pollution of the near-shore areas. How does this agricultural runoff affect coastal areas and coastal resources, such as commercial fisheries?

THEORY OF THE BIOSPHERE

Every science is based on theory and on models of how the object of study works. Theories about the biosphere have been based mainly on eighteenth- and nineteenth-century mathematics and physics, and most models treat the biosphere as a conventional mechanical system. We now realize that such concepts are inadequate. The biosphere is a dynamic, interactive system whose understanding requires new theoretical approaches. We need a new body of theory that

- provides a framework for understanding the influence of life on the cycling of chemical elements in the biosphere and on large-scale climatic changes,
- accounts for the long-term persistence of life and the role played by the diversity of species, and
- helps us understand the interrelationships of different species in the biosphere.

There have been many major changes in life and in the biosphere throughout Earth's history. Global changes in life have had effects on the air, waters, and land. The evolution of green plants, which led to the high concentration of oxygen in the atmosphere, is the most famous example. In turn, large-scale changes in the environment have affected life. We know about some of these changes, such as the great periods of continental glaciation and the extinction of the dinosaurs, but we cannot explain their causes. Theory needs experiments, and we can view Earth's history as a series of experiments with the biosphere; our attempts to explain them can form a basis for our theory and our understanding.

EXAMPLES OF FUNDAMENTAL SCIENTIFIC QUESTIONS

Evolution of Earth's Chemistry

What has led to the present chemical makeup of the atmosphere? The answer involves an understanding of astrochemistry, plate tectonics, atmospheric and ocean chemistry, and the interplay between these and life.

Evolution of Species

How have the number and kinds of species changed over time and what accounts for these changes? What role do various kinds of environmental fluctuations play in sustaining life on Earth?

Chemical Stabilities in the Biosphere

What factors lead to the amount of a chemical in some major unit of the biosphere (such as the atmosphere or a part of the ocean) remaining stable over a given period of time?

EXAMPLES OF APPLIED QUESTIONS

The Role of Change in Conservation

Classical ecological theory suggests that constant conditions lead to the greatest diversity of species and the forms of life most likely to persist. Recent theory suggests the opposite: that many species require change for their survival. We need a theoretical framework within which we can analyze the importance of change and disturbance in the conservation of species.

Biological Effects on Mineral Production

To what extent is a specific economically important mineral ore the product of biological processes? Determining this may help in locating economically viable sources of that mineral.

Famine

Frequent drought in Africa and elsewhere has led to major famines in recent years. What causes the drought remains unclear. Is it simply due to natural variations in climate? Or are distant human activities, such as in the cutting of tropical rain forests or the loss of forests from acid rainfall, changing climate globally, and thereby increasing the frequency of drought in Africa? Is the local desertification in Africa a self-enhancing process—the less vegetation, the less water evaporation, and the less local rainfall? Without an understanding of the underlying causes of drought, we are handicapped in our ability to respond.

CURRENT STATE OF THE BIOSPHERE

For any aspect of our life-support system, we need to know what the current condition is (for example, how much acid rain is descending on northern Europe, and how large an area of forest is under it?) and how the current condition is changing (Are the forests of Europe increasing or decreasing? At what rate? What changes in species are occurring with them?).

Despite what the detailed maps in your atlas would lead you to believe, we know very little about the quantity of life and its distribution on the Earth. In fact, almost none of the existing information has any statistical validity. The activities of GERO would include determining the current state of the biosphere and operating under the caveat that current conditions be investigated only in terms of a specific scientific or applied question; otherwise, the answer remains ambiguous, unapproachable, and unending.

Life on Earth is distributed broadly in various kinds of ecosystems. A type of ecosystem, such as grassland or tropical rain forest, is called a biome. Research on the current state of the biosphere will be done by major biomes or geographic regions.

EXAMPLES OF FUNDAMENTAL SCIENTIFIC QUESTIONS

Mapping Major Biomes

What geographic areas are covered by the major biomes? How can we make a map of these biomes that is useful for addressing a specific scientific question?

The Distribution of Life on Earth

How much organic material exists and how is it distributed over the Earth? What is the rate of change in the amount of organic matter?

The Distribution of Ocean Algae

The amount of algae in the oceans changes rapidly in time and space. What is the annual pattern of algae abundance and production worldwide?

EXAMPLES OF APPLIED QUESTIONS

Current Status of Carbon Production

How much carbon is stored in land vegetation at the present time? Is this amount increasing or decreasing? The biological reservoir of carbon has far-reaching implications for both the global climate and the persistence of a variety of terrestrial life forms.

Monitoring the Sources of Methane

What is the current level of biological production of methane? Major biological sources of methane production include marshland species, termites, and ruminant mammals. How is the rate of this methane production changing on a global basis? What does this imply for global climate and other aspects of atmospheric chemistry?

Measuring the Forests

How much commercial timber remains worldwide? What do changes in this level imply for climate, for generation of greenhouse gases like methane and carbon, and for the persistence of certain high-latitude life forms?

APPENDIX
III ACKNOWLEDGMENTS

The authors would like to acknowledge the National Science Foundation, the NASA Life Science and Planetary Sciences office, the Lounsbery Foundation, the University of California at Santa Barbara, the Commonwealth Book Fund, the Boston University Graduate School, and the Botany Department, University of Massachusetts at Amherst in support of this work.

We are grateful to Geraldine Kline and Caroline Lupfer for excellent manuscript preparation, and to Zachary Margulis for index programming. The index was compiled with the assistance of Theresa Chan, Kristine Hyon, and Rae Wallhausser. The work of Christie Lyons and J. Steven Alexander in preparation of Figures II–2 and i–I, respectively, and of Professors Stjepko Golubic and Ricardo Guerrero in Table AI-1 is acknowledged. We would also like to thank Rae Wallhausser and Wells Wilkinson for their assistance with figures and tables, as well as Gail Fleischaker, Gregory Hinkle, Joni Hopkins, Lorraine Olendzenski, and Dorion Sagan for many aspects of manuscript preparation.

INDEX

Italic page numbers refer to illustrations. Genera and species are italicized, the genus name capitalized.

187